Link Adaptation
for Relay-Based
Cellular Networks

RIVER PUBLISHERS SERIES IN COMMUNICATIONS

Volume 7

Consulting Series Editors

MARINA RUGGIERI
University of Roma "Tor Vergata"
Italy

HOMAYOUN NIKOOKAR
Delft University of Technology
The Netherlands

This series focuses on communications science and technology. This includes the theory and use of systems involving all terminals, computers, and information processors; wired and wireless networks; and network layouts, procontentsols, architectures, and implementations.

Furthermore, developments toward new market demands in systems, products, and technologies such as personal communications services, multimedia systems, enterprise networks, and optical communications systems.

- Wireless Communications
- Networks
- Security
- Antennas & Propagation
- Microwaves
- Software Defined Radio

For a list of other books in this series, see final page.

Link Adaptation
for Relay-Based
Cellular Networks

Başak Can

River Publishers

Aalborg

ISBN 978-87-92329-30-1 (hardback)

Published, sold and distributed by:
River Publishers
P.O. Box 1657
Algade 42
9000 Aalborg
Denmark

Tel.: +45369953197
www.riverpublishers.com

To my husband Özgür Oyman

To my parents Leman & Mustafa Can and my brother Ceyhun Can

To the peaceful communication in the world and in the space

Peace in the country, peace in the world.
Mustafa Kemal Atatürk

To live like a tree, single and free
and brotherly like the trees of a forest ...
Nazım Hikmet Ran, from his poem named "Invitation"

Don't quit when the tide is lowest,
For its just about to turn.
Don't quit over doubts and questions,
For there's something you may learn.
Don't quit when the night is darkest,
For it's just a while 'til dawn;
Don't quit when you have run the farthest
For the race is almost won.
Don't quit when the hill is steepest,
For your goal is almost nigh;
Don't quit, for you're not a failure
Until you fail to try.
Jill Wolf

Contents

Preface

This book focuses on the implementation of various link adaptation methods in OFDM(A) (Orthogonal Frequency Division Multiplexing/Multiple Access)-Time Division Duplex (TDD)-based two-hop cellular networks. The analysis and design consider infrastructure-based relays. New link adaptive transmission methods which dynamically select the best coding, modulation, forwarding, relaying mechanisms and the packet size have been designed and evaluated for such networks.

In this book, a link adaptation and selection method for the links constituting an Orthogonal Frequency Division Multiplexing (OFDM)-based wireless relay network is proposed. The proposed link adaptation and selection method selects the forwarding, modulation and channel coding schemes providing the highest end-to-end throughput and decides whether to use the relay or not. The link adaptation and selection is done for each sub-channel-based on instantaneous SINR conditions in the source-to-destination, source-to-relay and relay-to-destination links. The considered forwarding schemes are Amplify and Forward (AF) and simple adaptive Decode and Forward (DF). Efficient Modulation and Coding Scheme (MCS) selection rules are provided for various relaying schemes. The proposed end-to-end link adaptation and selection method ensures that the end-to-end throughput is always larger than or equal to that of transmissions without relay and non-adaptive relayed transmissions. The evaluations show that over the region where relaying improves the end-to-end throughput, the DF scheme provides significant throughput gain over the AF scheme provided that the error propagation is avoided via error detection techniques. A frame structure to enable the proposed link adaptation and selection method for OFDMA (Orthogonal Frequency Division Multiple Access)-TDD-based relay networks is provided. This frame structure can be used for the emerging IEEE 802.16j standard.

This book provides a channel adaptive scheduler which considers the multiplexing loss caused by the two-phase nature of wireless relaying. The scheduler dynamically schedules the users on the frequency-time radio resources with efficient MCSs selected by Adaptive Modulation and Coding (AMC). Relaying is used only if it can provide throughput enhancement. It is assumed that the scheduler has the instantaneous channel state information of the source-to-destination and relay-to-destination links. Further in this book, the guidelines for efficient deployment of infrastructure-based relay terminals are given.

For the emerging IEEE 802.16j standard, the system level performance of various cooperative diversity schemes has been investigated with the scheduler developed and the relays efficiently deployed in the cell. This investigation shows that, a simple cooperative diversity scheme that dynamically chooses the best scheme among direct transmission and two-hop conventional relaying is a promising choice when compared to various more complex cooperative diversity schemes.

The end-to-end maximum achievable rate of various relaying schemes such as Cooperative-Multiple Input Multiple Output (MIMO), Cooperative-Single Input Multiple Output (SIMO), Cooperative-Multiple Input Single Output (MISO) and conventional relaying is analyzed. Comparative analysis of these schemes is presented for both AF and DF-based relaying. The comparisons from information theoretic view show that cooperative-MIMO and cooperative-MISO schemes can outperform each other. The cooperative-MISO scheme can outperform cooperative-SIMO scheme but cooperative-SIMO scheme cannot outperform cooperative-MISO scheme. If the SINR condition in the relay links is much larger than that of source-to-destination link, all the relaying schemes perform the same. DF-based relaying can provide significant gain in transmission rate as compared to AF-based relaying. A fully adaptive relaying scheme which dynamically uses the best relaying scheme and decides whether to relay or not is analyzed. The performance of such adaptive relaying scheme is compared to each one of the relaying schemes which dynamically decides to relay or not. As the information theoretic investigations show, a simple scheme which dynamically decides the best scheme among conventional relaying and direct transmission can perform similar to more complex adaptive relaying schemes. This conclusion agrees with the conclusion drawn from practical implementation with AMC. Hence, by using this simple scheme, substantial savings from complexity can be achieved at the mobile station (MS).

In this book, a hop adaptive Medium Access Control (MAC)-Protocol Data Unit (PDU) size optimization is proposed for wireless relay networks. With this proposal, the MAC-PDU size in different hops can be different and it is optimized based on the channel condition of each hop. Such optimization improves the end-to-end goodput via MAC-PDU size optimization. The proposal further reduces the total overhead transmitted in the end-to-end path via transmitting longer length packets in the potentially robust Base Station (BS)-to-Relay Station (RS) links.

Various synchronization issues for OFDM(A)-based wireless relay networks have been analyzed. The time and carrier frequency offset issues have been addressed. A new method for relieving the effects of time offset has been proposed. The analysis shows that, the time and carrier frequency offset issues are not problematic for infrastructure-based relaying. Hardware complexity of various cooperative diversity scheme implementations are investigated for the MS. The investigations show a significant complexity increase with the cooperative diversity schemes which require coherent signal combining at the MS. Hence, coherent signal combining at the MS should be used only if there is a throughput gain.

Acknowledgments

This book is based on my Ph.D. thesis that was conducted at Aalborg University, Denmark and includes minor revision on my Ph.D. thesis.

I would like to thank my parents Mustafa & Leman Can and my brother Ceyhun Can who have supported me throughout my life. I appreciate their patience when I was away from home to conduct my Ph.D. studies.

I would like to thank Professor Ramjee Prasad for his support and guidance throughout my Ph.D. studies and for his recommendation on writing this book. I would like to thank Professor Hiroyuki Yomo and Professor Elisabeth De Carvalho for their guidance through my Ph.D. studies.

I would like to thank my husband Özgür Oyman for his support in writing this book. I am grateful to my studies presented herein in bringing us together.

My Ph.D. thesis was jointly sponsored by the following organizations: Aalborg University, Denmark, Telecommunication R&D Center, Samsung Electronics Co. Ltd., Suwon, Republic of Korea, Carleton University, Canada, and Intel Corporation, U.S.A. I would like thank these organizations for sponsoring my thesis.

I would like to thank Professor Halim Yanikomeroglu for inviting me to Carleton University, Canada for research collaboration at the Systems and Computer Engineering Department. I would like to thank him not only for his valuable guidance and comments for my thesis but also for his support on my career and success. I would like to thank Furuzan Atay Onat for her support and comments on the work presented in Chapter 4.

I would like to thank Professor Peter Koch and Professor Nail Akar for their guidance on my admission to Aalborg University. I would like to thank Professor Ole Olsen for his support on my Ph.D. studies.

I would like to thank Euntaek Lim, Marcos Katz, Dong Seek Park, Hokyu Choi, Youngkwon Cho, Sivanesan Kathiravetpillai and David Mazzarese who have been working at Samsung Electronics, Co. Ltd., Suwon, Korea for their comments on the work presented in my thesis.

I would like to thank Maciej Portalski, Hugo S. Lebreton, Yannick Le Moullec and M. Imadur Rahman for their collaboration on investigating the hardware implementation aspects of OFDMA-based wireless relaying. A part of the outcome of this collaboration is presented in Section 6.2. The further results and more details of this investigation can be found in [1, 2] and Maciej's Master of Science thesis titled "Implementation Aspects of Fixed Relay Station Design in OFDM(A) Based Wireless Relay Networks". I would like to thank Jesper M. Kristensen in being a sensor and giving valuable comments for Maciej's master of science defense.

I would like to relate my special thanks to Rath Vannithamby, Hyunjeong Hannah Lee and Ali Taha Koç for their supervision and guidance on various sections presented in my thesis. I would like to thank Brian J. Sublett, Xiaoshu Qian, Yi Hsuan, Atul Salvekar, Jong-Kae Fwu, Xuemei Ouyang, Baraa Al Dabagh, Jinyong Lee, Yongfang Guo, Balvinder S. Bisla, Harish Balasubramaniam and Kevin Klesenski for their support during my work at Intel Corporation within Wireless Technology and Products Group. I have shared the hard working times with them in the harmony of joy.

I would like to thank Yannick Le Moullec, Persefoni Kyritsi, M. Imadur Rahman, Simone Frattasi, Hanane Fathi, Huan C. Nguyen, Nicola Marchetti, Megumi Kaneko, Yasushi Takatori, Rasmus Løvenstein Olsen, Suvra Sekhar Das, Yaoda Liu, Elvis Bottega, Hamid Saaedi and Abdulkareem Adinoyi for their valuable comments and support on my Ph.D. studies. I would like to further thank Yannick Le Moullec and Ernestina Cianca for giving me comments on my thesis.

I would like to thank Jane Yee for her endless support and friendship during the times when I was finalizing my thesis. I would like to thank my music teachers Mahir Güçlü and Ulaş Acar for the music classes I took from them during my studies. Special thanks to Mahir Güçlü for a mini-concert that we have given together in Canada. These music activities have cheered me up during my Ph.D. studies.

I would like to thank all the people I met during these years, with whom I shared hard working times as well as fun times. As a former member, I would like to thank the members of Aalborg Volleyball Club (Aalborg VK74) whom cheered me up during my Ph.D. studies.

In memory of my grandparents and Hanife Çakır,

Başak Can
İzmir, Türkiye (Turkey)
May 2009

List of Figures

List of Tables

Acronyms

2G	Second Generation
3G	Third Generation
4G	Fourth Generation
AAF	Alamouti Amplify and Forward
ACK	ACKnowledgment
AdDF	Adaptive Decode and Forward
AF	Amplify and Forward
AMC	Adaptive Modulation and Coding
ARQ	Automatic Repeat Request
ATM	Asynchronous Transfer Mode
AWGN	Additive White Gaussian Noise
BEP	Bit Error Probability
BER	Bit Error Rate
BPSK	Binary Phase Shift Keying
BS	Base Station
BSC	Binary Symmetric Channel
BSN	Block Sequence Number
CFO	Carrier Frequency Offset
CQICH	Channel Quality Indication CHannel
CRC	Cyclic Redundancy Check
CSI	Channel State Information
CP	Cyclic Prefix
DF	Decode and Forward
DFT	Discrete Fourier Transform
DL	Down-Link
DSP	Digital Signal Processing
EIRP	Effective Isotropic Radiated Power
FC	Fragmentation Control
FCH	Frame Control Header
FEC	Forward Error Correction
FFT	Fast Fourier Transform

FPGA	Field Programmable Gate Array
FSH	Fragmentation Subheader
FSN	Fragment Sequence Number
GI	Guard Interval
HARQ	Hybrid ARQ
ICI	Inter Carrier Interference
IDFT	Inverse Discrete Time Fourier Transform
IFFT	Inverse Fast Fourier Transform
IPR	Intellectual Property Right
ISI	Inter Symbol Interference
LO	Local Oscillator
LOS	Line-of-Sight
MAC	Medium Access Control
MAN	Metropolitan Area Network
MAP	Map
MCS	Modulation and Coding Scheme
MIMO	Multiple Input Multiple Output
MISO	Multiple Input Single Output
ML	Maximum Likelihood
MRC	Maximum Ratio Combining
MS	Mobile Station
NLOS	Non-Line-of-Sight
OFDM	Orthogonal Frequency Division Multiplexing
OFDMA	Orthogonal Frequency Division Multiple Access
OFDM(A)	Orthogonal Frequency Division Multiplexing/Multiple Access
PDU	Protocol Data Unit
PER	Packet Error Rate
PFS	Proportional Fair Scheduler
PHY	PHYsical
PSH	Packing Subheader
PSK	Phase Shift Keying
QAM	Quadrature Amplitude Modulation
QoS	Quality of Service
QPSK	Quadrature Phase Shift Keying
RF	Radio Frequency
RHS	Right-Hand Side
RMS	Root Mean Square
RS	Relay Station
Rx	Receive

SDU	Service Data Unit
SEP	Symbol Error Probability
SIMO	Single Input Multiple Output
SINR	Signal-to-Interference-Plus-Noise-Ratio
SISO	Single Input Single Output
SNR	Signal-to-Noise-Ratio
STD	Space Time Decoding
STDF	Space Time Decode and Forward
TDD	Time Division Duplex
TDMA	Time Division Multiple Access
Tx	Transmit
UB	Upper Bound
UL	Uplink
VAA	Virtual Antenna Array
WAN	Wide Area Network
w/o	without
ZMCSCG	Zero Mean Circularly Symmetric Complex Gaussian

1

Introduction

As of year 2008, wireless cellular networks have existed for decades. In their infancy phase, such networks were deployed mainly for voice communications and have been referred to as Second Generation (2G) wireless networks. The continuing research for the further development of wireless networks has enabled multi-media transmissions to the end users. Such developments created the so called Third Generation (3G) wireless networks. The 3G systems such as IMT-2000 can provide 144 kb/s for high mobility, 384 kb/s for nomadic and 2 Mb/s for low mobility users [7]. Nowadays, the users would like to have mobile multi-media services as good as wired broadband networks. With the 3G systems, meeting such a demand is expensive and not possible [7]. Currently, tremendous research activities take place for the development of systems beyond the capabilities of 3G systems. The Fourth Generation (4G) systems are being developed with this purpose. The development of the 4G systems targets to provide ubiquitous high data rate coverage at lower costs to both the system operators and the end users. The 4G systems target to provide a data rate of 100 Mb/s to mobile users and a data rate of 1 Gb/s/Hz to nomadic users [7]. Hence, efficient mechanisms to improve the throughput should be developed rather than focusing on improving Bit Error Rate (BER) only.

The 4G systems will take into account the heterogeneity of the wireless terminals which would like to access to the network. Furthermore, the 4G networks target to provide a better Quality of Service (QoS) as compared to their previous counterparts. In summary, the 4G networks will provide wireless communications anytime, anywhere and anyhow [8].

The achievement of the targeted goals for the successful implementation of the 4G networks is challenging. Such achievement is not practical via Single Input Single Output (SISO) links due to the limited spectral efficiencies that they offer [9]. To achieve the targeted goals, the development of new transmission and reception strategies are required. Multiple-antenna techniques, which are referred to as Multiple Input Multiple Output (MIMO) techniques, are one of these strategies which lead to an increase in the capacity of the system.

MIMO communication with co-located multiple antennas at a given terminal provides tremendous advantages with some challenges. It offers spatial diversity, spatial multiplexing, beam-forming, space division multiplexing, interference suppression, etc. [9–12]. In rich scattering environments, MIMO communication provides much better spectral efficiencies as compared to that of SISO links. Therefore the MIMO techniques with the co-located antennas are promising for increasing the spectral efficiency of the wireless networks. The capacity increases linearly with the number of transmit antennas as long as the number of the receive antennas is greater than or equal to the number of the transmit antennas [9]. Spatial diversity techniques provide the receiver multiple signatures of the same transmitted signal. Each signature is referred to as a diversity branch. If the number of independent signatures of the same transmitted signal increases at the receiver, then the probability that all the diversity branches will be in fade reduces. This leads to an increase in the reliability and hence the capacity of the link between the transmitter and the receiver [10].

There are some challenges which arise from the MIMO communication. The performance improvements via the MIMO techniques cannot be always achieved. The performance of the MIMO communication depends strongly on:

- the antenna element numbers at each terminal,
- the scattering environment,
- the inter-element spacing,
- the presence of Line-of-Sight LOS component.

The latter three directly influences the spatial fading correlations between Transmit (Tx) and Receive (Rx) antennas [10]. In order to achieve the benefits of the MIMO communication, the spatial correlation and mutual antenna coupling between the co-located antennas should be sufficiently low [10]. This is especially crucial for the spatial diversity and spatial multiplexing schemes. This necessitates an inter-element spacing of $k(\lambda/2)$, where λ represents the wavelength and $k \in \{1, 2, 3, \ldots\}$ [13]. The emerging 4G wireless communication systems such as IEEE 802.16e (which is also referred to as mobile-WiMAX) based systems operate at the carrier frequencies

2.3, 2.5, 3.3 and 3.5 GHz in the licensed spectrum allocations [3]. A carrier frequency of 2.5 GHz corresponds to a wavelength of 12 cm. For such carrier frequencies, it may not be possible to put more than two antennas at small hand-held terminals and still fully achieve the benefits of the MIMO communication [10].

The MIMO techniques require multiple Digital Signal Processing (DSP) and Radio Frequency (RF) units at the terminals. Multiple RF chains bring costs in terms of the size, power consumption, hardware and hence price. As an example consider an OFDM-MIMO system. To achieve MIMO-OFDM advantages, one needs to have multiple parallel Orthogonal Frequency Division Multiplexing (OFDM) modulators/demodulators and RF modules in the transceiver [10]. The OFDM-based systems such as IEEE 802.16e systems will support up to 2048 sub-carriers. For such standards, it is computationally complex to perform multiple simultaneous OFDM modulations/demodulations at the mobile terminals with the current limited battery power and processing capability. Consequently, even if the co-located antennas are spatially uncorrelated, it is still complex to achieve the benefits of the MIMO-OFDM communication especially for the mobile terminals with the limited size, power and processing capability. To remedy these challenges, the advantages of the multi-antenna techniques can be achieved via an integrated approach based on cooperative wireless communications. In the following section, the motivation for the cooperative wireless communications is presented.

1.1 Introduction to and Motivation for Cooperative Communications

Cooperative wireless communication has become an active area of research since 1998 where each terminal acts as an information source as well as a relay. The relay terminals can be a mobile terminal in the network, e.g., a mobile subscriber unit. Such terminals are referred to as mobile relays. The relay terminals can be deployed by a system operator to be exclusively used for relaying. Such terminals are referred to as fixed or infrastructure-based relay terminals. Cooperative communications create a Virtual Antenna Array (VAA) in the system without the need of multiple antennas, multiple RF modules and multiple modulation units co-located at a given terminal. With such VAA, many advantages provided by MIMO techniques can be achieved [5, 14–18]. A VAA can be referred to as cooperative multi-antennas. In coo-

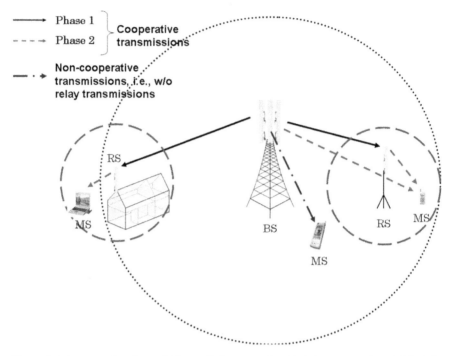

Figure 1.1 An example of a multi-hop cellular network with infrastructure-based RSs. The source is a BS and the destinations are MSs.

perative wireless communication, the information from a source is forwarded by one or several RS(s) to a destination. Hence, cooperative communication is arranged in several transmission phases or hops and referred to as multi-hop communications as well. It can be applicable in a wide variety of wireless network settings. For example it can be implemented in cellular and ad–hoc networks. Any kind of wireless network using relay nodes can be referred to as multi-hop or wireless relay networks. If relaying is used in a cellular network, then such a network is referred to as multi-hop cellular network. If infrastructure-based relay terminals are deployed in a multi-hop cellular network, such a network is referred to as infrastructure-based wireless relay network. An example of an infrastructure-based wireless relay network is depicted in Figure 1.1. With the current technology, the wireless terminals cannot transmit and receive using the same radio resource [6, 15, 19]. There-fore, the cooperative communication in wireless relay networks is organized in several phases. A phase for the reception at the RS from the source terminal or from the previous intermediate node is needed. For the case of multi-hop transmission, the intermediate node refers to one of the relay(s) in the end-to-end path. Another phase for the transmission from the RS, i.e. forwarding, to the destination terminal or the next intermediate node is needed. This multi-phase structure results in a multiplexing loss due to the need for additional radio resources for forwarding. The RSs use a forwarding method for relaying. The different transmission sequences of the relay and the source terminals lead to a different relaying scheme. In Figure 1.1, two different relaying schemes are depicted. One of these relaying schemes provides the MS signals received both form the BS and the RS. Such a scheme can be realized in regions where the mobiles are within the coverage of both the BS and the RS. Whereas with the other relaying scheme, the MS needs to rely on the signals received solely from the RS. This relaying scheme is referred to as conventional relaying. Such a scheme can be beneficial over the regions where the MSs are out of the coverage of the BS while they are within the coverage area of the relay.

In order to reduce the deployment costs of cellular networks via wireless relaying, the infrastructure-based wireless relay terminals should be simple devices as compared to the base stations. They may be controlled by a base station and should not necessarily posses Medium Access Control (MAC) functionalities. They should not require wired connection to the backbone and should be comprised of simple antenna and Rx/Tx chains. They should transmit at lower powers as compared to a BS. This way, infrastructure-based wireless relay terminals can be low cost terminals [20].

The OFDMA-based IEEE 802.16-2004 standard is developed for fixed broadband wireless applications in Metropolitan Area Networks (MANs) [21]. It does not support user mobility. The Orthogonal Frequency Division Multiple Access (OFDMA)-based IEEE 802.16e standard is developed for providing broadband coverage for mobile users in single hop wireless MANs [3]. It can provide coverage for both LOS and Non-Line-of-Sight (NLOS) conditions. It enables QoS support and various adaptive transmission technologies such as Adaptive Modulation and Coding (AMC), adaptive sub-channelization, power control, adaptive multiple antenna techniques, effective user scheduling mechanisms, etc. In NLOS conditions, it can provide a coverage range from the BS as much as 8 km [22]. This standard is envisioned as one of the first 4G technology being deployed in various regions in the world. The emerging wireless transmission standard IEEE 802.16j will increase the coverage area and the aggregate throughput of the IEEE 802.16e-based wireless networks by enabling multi-hop transmissions. By designing relatively cheap RSs as compared to a BS, it targets to reduce the deployment costs of the IEEE 802.16e-based cellular networks. The IEEE 802.16j group [23] is working on the successful implementation of cellular multi-hop wireless networks.

The following advantages are offered by wireless relay networks. The destination terminal can obtain the benefits of cooperative diversity (akin to spatial diversity) via appropriately combining the received signals from the source and relay terminals. The cooperative diversity leads to an increase in the throughput and hence the coverage area of the system. As a consequence, the outage probability and the BER of the communication between the source and the destination terminals can be reduced. Mobiles become less susceptible to the variations in the channel.

In MIMO communication, shadowing will affect all the antennas in a similar way as they are co-located. With the cooperative communication, the mobiles are less susceptible to the shadowing effects as the antennas are not co-located. For example, if the source-to-destination ($S \rightarrow D$) link is in a deep shadow, the source-relay-destination ($S \rightarrow R \rightarrow D$) link might not be in a shadow. This brings the macro diversity benefits. With the aid of the added reliability in the end-to-end communication, the terminals can transmit using higher rate modulation modes. This leads to an increase in the total transmission rate and hence reduces the multiplexing loss caused by the need for multi-hop transmissions [14]. The cooperative terminals can transmit at lower powers while maintaining the link quality, coverage and/or achievable rate pair at the same level obtained without cooperative strategies [14]. This extends the battery life of the mobiles and reduces the interference in the system.

Unlike MIMO communication with co-located antennas, the spatial fading correlations in a virtual antenna array is expected to be very low as the antennas are not co-located. Hence, the benefits of spatial multiplexing and spatial diversity offered by the VAA can be achieved fully.

The multi-hop cellular networks have the potential of enhancing the system coverage and throughput as compared to their single hop counterparts [18]. Such networks have been considered as the promising candidates for the successful deployment of next generation wireless networks such as 4G [20, 24]. One of the main challenges of the 4G communications is the increased path loss due to the usage of higher carrier frequency bands as compared to that of 2G and 3G networks. Therefore, it is difficult to provide high data rate coverage over long distances. To remedy this, one might reduce the cell size by deploying more BSs to cover a certain geographical area. However, this drastically increases the deployment cost hence contradicts the target of the 4G networks. Relay terminals deployed at strategic locations in a cellular wireless communication network can remedy this problem. If the source-to-relay ($S \rightarrow R$) and the relay-to-destination ($R \rightarrow D$) distances are smaller than the source-to-destination ($S \rightarrow D$) distance, then the distance dependent path loss in the $S \rightarrow R$ and the $R \rightarrow D$ links are smaller than that of $S \rightarrow D$ link. In such a case, the destination terminal can further benefit from the reduced end-to-end path loss. As a consequence, the high data rate coverage area can be increased without the need for increasing the total number of serving base stations. This way, the deployment costs to cover a certain geographical area can be reduced via using low cost relay terminals.

The motivation in this book is based on the afore-mentioned advantages of the cooperative wireless communication, which can remedy the challenges in conventional MIMO communication. The problem statement and the book outline is presented in the following section.

1.2 Problem Statement and Book Outline

There are various issues that need careful design for the successful implementation of multi-hop cellular networks. These include but are not limited to the following:

- functional division between the BS and the RSs,
- transmission sequence design of the source and relay terminals,
- scheduling and radio resource management,
- power control,
- frequency planning,
- efficient deployment of the relay stations,
- maintaining user transparent functionality,
- synchronization,
- minimization of control information overhead and
- hardware implementation aspects.

The theoretical analysis of the wireless relay networks in frequency flat environments is well covered in the literature [5, 25–27]. However, the practical design of efficient AMC, forwarding and relaying mechanisms as well as consideration of the end-to-end performance and frequency selective environments are not well covered in the literature.

This book focuses on the design of various PHYsical (PHY) and MAC layer mechanisms for the efficient implementation of Orthogonal Frequency Division Multiplexing/Multiple Access (OFDM(A))-based two-hop cellular

networks. New link adaptive transmission methods which dynamically select the best channel coding, modulation, forwarding, relaying mechanisms
and the packet size have been designed for low mobility users. The analysis
and design are provided with the following main assumptions. The source,
relay and destination terminals have one antenna for transmitting and receiving. Down-Link (DL) transmissions in a single cell with zero intra-cell
interference has been considered.

In this book, the following problems are identified and solutions to them
are proposed:

- **Determination of the conditions where relaying is beneficial:**
  Cooperative communication suffers from multiplexing loss. Therefore,
  relaying cannot always improve the end-to-end performance as
  compared to the without (w/o) relay transmissions. To this end, relaying
  should be introduced only when there is end-to-end performance
  improvement. Determination of the conditions where relaying should
  be used has not been thoroughly analyzed in the literature. The up to
  date literature has either focused on only Signal-to-Noise-Ratio (SNR)
  conditions in the links or the information theoretic performance with
  a specific relaying mechanism. However, only SNR conditions cannot
  provide information on the multiplexing-loss inherent in relaying.
  The information theoretic analysis can provide analysis on the system
  capacity and fundamental benefits of relaying. However, it is not enough
  to provide solutions in a practical system setup. Therefore, this book
  provides analytical derivations, look-up tables and extensive analysis
  on the conditions where relaying should be used in a practical system
  setup. This analysis is not limited to a specific relaying mechanism or
  MCS. Rather, it includes the analysis for various relaying mechanisms
  with AMC. Such an analysis is presented in Chapters 2, 3 and 4.

- **Comparative performance analysis of the main forwarding
  schemes and the design of adaptive forwarding:** There are various
  forwarding mechanisms used for wireless relaying. AF and DF are
  two main forwarding schemes which have been extensively used in
  the cooperative wireless communications. The relative performance
  with either of these forwarding mechanisms depends strongly on the
  relaying scheme, the transmission rate and the channel conditions.
  Hence, in order to determine the beneficial forwarding scheme in
  terms of spectral efficiency or better reliability, one needs to compare

the end-to-end performance with each of the forwarding schemes. The analysis and conclusions should not be limited to one relaying scheme and one transmission rate. Furthermore, the comparative analysis and conclusions should be provided over the region where relaying improves the spectral efficiency or the reliability. This way, the beneficial forwarding mechanism can be used once relaying is decided. These issues are analyzed in Chapters 3 and 4 from both implementation and information theoretic point of view.

- **The design of the link adaptation methods for wireless relay networks:** In conventional single-hop networks, the link adaptation mechanisms deal with the channel conditions between only the source and the destination terminals. However, in wireless relay networks, the channel conditions between the source and destination, as well as between the source and relay, and the relay and destination should be considered. There is a need for the design of link adaptation mechanisms which consider the end-to-end optimization in wireless relay networks. Such mechanisms are developed in Chapter 3 for various forwarding and cooperative diversity schemes.

- **The design of the frame structure to accommodate multi-hop communications:** In conventional single-hop cellular networks, the frame structure is designed for information flow between the end users and the base station. In multi-hop cellular networks, the information flow should be controlled not only for the source-user transmissions but also for the relay-user and source-relay transmissions. The functional division between the relays and the base station should be clearly mapped on the frame structure. The control signalling should be provided not only to the users but also to the relays. In order to use the link adaptive transmission mechanisms developed in this book, a frame structure is developed. This frame structure is based on the IEEE 802.16j standard. However, it shows where the BS and the RS will transmit based on the radio resource allocation. This frame structure can be used in the DL of two hop cellular networks. It is presented in Chapters 3 and 4.

- **The efficient deployment of the infrastructure-based relay terminals:** An important issue in infrastructure-based relay networks is the deployment of the relays on strategic positions in the cell. To efficiently extract the benefits of wireless relaying, the relays should neither be deployed at very close distances to the BS nor at very far distances to the BS. The link quality in the BS to RS ($BS \rightarrow RS$) links should be maintained while providing efficient coverage from the relays. The guidelines for the efficient deployment of the infrastructure-based relay terminals in IEEE 802.16j-based two-hop cellular networks is provided in Chapter 4.

- **Radio resource allocation and user scheduling in wireless relay networks:** The radio resource allocation and user scheduling in wireless relay networks differ from conventional single-hop networks. This is due to the fact that wireless relaying involves multiple links for transmission towards a given user. The end-to-end performance should be considered while taking into account all the link conditions in the end-to-end path. The users should be scheduled with the most efficient scheme among cooperative and non-cooperative transmissions. Such radio resource allocation and user scheduling are developed in Chapter 4.

- **Comparative performance analysis of various relaying schemes and the design of adaptive relaying:** Different relaying schemes result in different cooperative multi-antenna schemes and hence can provide different cooperative diversity schemes. Each of these cooperative multi-antenna schemes achieve different end-to-end throughput. Each of them can, to a certain extend, compensate for the multiplexing loss. There is not a unique relaying scheme which can outperform all the other schemes. The relative performance depends on the channel conditions and the end-to-end net throughput offered by each relaying scheme. To determine the most efficient relaying scheme for given channel conditions, analytical derivations and look-up tables should be developed. The system level comparative performance evaluations of various relaying schemes should be provided. Adaptive relaying refers to the scheme where the relaying scheme which provides the highest spectral efficiency is selected for given channel conditions. Whether such adaptive relaying is beneficial or not should be investigated. These issues are investigated in Chapter 4 from an implementation point of view as well as from an information theoretic point of view. For investigations from an implementation point of view, the relative performance of various link adaptive cooperative diversity schemes has been analyzed with AMC. These investigations provide guidelines for the design of IEEE 802.16j-based wireless relay networks. The investigations are provided in terms of average end-to-end throughput and the coverage area of a cell operating based on IEEE 802.16j and IEEE 802.16e standards.

- **Packet size optimization in wireless relay networks:** Larger packets (in terms of total number of bits per packet) transmitted in wireless medium encounters a larger Packet Error Rate (PER). On the other hand, increasing the packet size reduces the significance of the overheads, e.g., headers and Cyclic Redundancy Check (CRC), carried in each packet. Smaller packets transmitted in wireless medium encounters a smaller PER. However, this results in an increase in the significance of the overheads carried in each packet and hence reduces the net throughput delivered to the upper layers. When the channel conditions, i.e., the SNR, are good enough to provide negligible a PER with a given MCS, then increasing the packet size will be beneficial. Vice versa, when the channel conditions are not good enough to provide acceptable PER with a given MCS, then decreasing the packet size may reduce the PER. In conclusion, the AMC should be accompanied with adaptive packet size selection per channel condition. The packet size selection/optimization in wireless relay networks should consider all the links constituting the end-to-end path and should be adjusted for each hop. Such new hop-adaptive packet size optimization is proposed, developed and evaluated in Chapter 5.

- **Implementation issues for wireless relay networks:** There are various implementation issues that need to be addressed for the successful implementation of wireless relay networks. One of these issues is synchronization. The relay stations should transmit on dedicated time slots and frequencies. Some relaying schemes necessitate simultaneous transmissions from the source and the relay terminals. This may lead to some synchronization problems such as time and frequency offset. Such problems are investigated and solutions are proposed in Section 6.1 for infrastructure-based two-hop cellular networks. The other crucial implementation issue is the hardware aspects of cooperative communications especially at the MSs where low complexity of operation is important. Cooperative diversity achieved in the form of coherent signal combining increases the complexity at the MS. Achievement of different cooperative diversity schemes results in different hardware complexity at the MS. Therefore, the complexity of each scheme should be investigated to determine the most practical cooperative diversity scheme for the low cost MSs. These issues are addressed in Section 6.2.

Chapter 2 gives the background on cooperative communications. Finally, Chapter 7 gives the overall conclusions and directions for the future work.

2

Background on Cooperative Communications and the Systems Considered*

2.1 Introduction

This chapter presents the background on cooperative communications and the systems considered. It provides the system model considered in this book and gives the motivation to use link adaptive transmission techniques for OFDM(A) based wireless relay networks. An RS using the AF scheme amplifies and forwards (without decoding) the signal which is received from the source [5]. For OFDM-TDD based wireless relay networks, the transceiver structure of the relay terminal using the AF scheme is depicted in Figure 2.1. With TDD operation, the two phases (one for reception at the relay and one for forwarding) are separated in time. With AF based relaying, these two time phases need to have equal duration as the BS and the RS transmit with the same constellation. The OFDM modulator consists of Inverse Fast Fourier Transform (IFFT), Cyclic Prefix (CP) insertion, parallel to serial converter and digital to analog converter blocks. The OFDM demodulator block consists of analog to digital converter, CP deletion, serial to parallel converter and FFT blocks. In the first phase, the RS using AF based relaying performs FFT operation and buffers the complex digital signal in each sub-carrier for subsequent transmission in the second phase. It amplifies the signal at each sub-carrier subject to its total average transmit power. To maintain its total average transmit power at a given value, the RS uses the CSI information of the $S \rightarrow R$ link. For forwarding in the second phase, the RS remodulates the signal at each sub-carrier with OFDM. The amplification and forward-

---
*The work presented in this chapter has been presented at the IEEE International Conference on Communications (ICC), İstanbul, Turkey, June 2006, and can be found in [28]. A patent on this work has been filed by the Korean Intellectual Patent Office and can be found in [29]. © 2006 IEEE. Reprinted with permission from *Proceedings of the IEEE International Conference on Communications (ICC), Istanbul, Turkey, June 2006.*

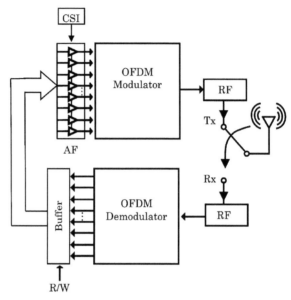

Figure 2.1 The transceiver structure of the relay terminal with AF based relaying in an OFDM-TDD based wireless relay network. The buffering and amplification is done at each sub-carrier in order to cope with frequency selective fading.

ing could be done without the need for OFDM demodulation. However in this case, link adaptation and selection cannot be done for each sub-channel. Such a case does not enable efficient use of radio resources and may cause instability.

For OFDM-TDD based wireless relay networks, the transceiver structure of the relay terminal using the DF scheme is depicted in Figure 2.2. An RS using the DF scheme demodulates, decodes, buffers, re-encodes, modulates and forwards the signal which is received from the source terminal [5]. The buffer is necessary to store the decoded bits received at each sub-carrier during the first phase. Hence, the memory size requirement of the buffer used for the DF based relaying will be less than that of AF based relaying. This is especially due to the fact that, with DF based relaying, the bits are stored instead of complex numbers. However, the DF based relaying is more power consuming at the relay terminal as it necessitates more processing as compared to AF based relaying [1].

The wireless relay networks with the two-hop communication involve three links: $S \rightarrow R$, $R \rightarrow D$ and $S \rightarrow D$. An example of a VAA is visualized

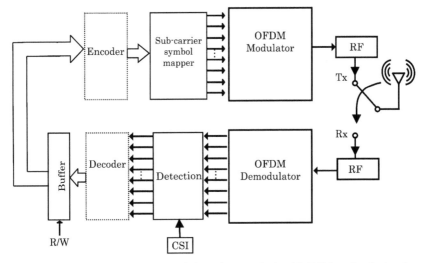

Figure 2.2 The transceiver structure of the relay terminal with DF based relaying in an OFDM-TDD based wireless relay network.

Table 2.1 Cooperative protocols leading to cooperative-MIMO, SIMO, MISO channels [5] and conventional relaying [6].

| Transmission | Transmission sequence leading to cooperative | | | Conventional |
Phase	MIMO	SIMO	MISO	Relaying
Phase-1	$S \to R, D$	$S \to R, D$	$S \to R$	$S \to R$
Phase-2	$S \to D, R \to D$	$R \to D$	$S \to D, R \to D$	$R \to D$

in Figure 2.3. It emulates a MIMO communication system and hence allows various MIMO techniques[1] to be used in a distributed manner.

For wireless relay networks, there are various cooperative multi-antenna schemes proposed in the literature. These schemes can provide cooperative transmit diversity and cooperative receive diversity. Relaying scheme refers to a certain relaying scheme which achieves a certain cooperative-multi-antenna-channel such as cooperative-MIMO, cooperative-Multiple Input Single Output (MISO), cooperative-Single Input Multiple Output (SIMO) and cooperative-SISO. The cooperative-SISO channel refers to the channel achieved with conventional relaying. The relaying schemes analyzed in this book are listed in Table 2.1. The terminology $A \to B$ represents the wireless transmission from terminal A to B. The term $A \to B, C$ repres-

---

[1] E.g., transmit and/or receive diversity and Space Time Coding (STC).

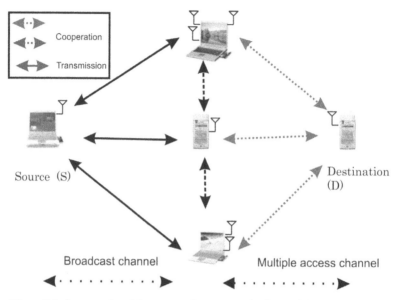

Figure 2.3 An example of the cooperative communication and the created VAA.

ents the case where terminal A is transmitting and both terminal B and C are receiving. The cooperative-multi-antenna-channels namely cooperative-MIMO, cooperative-SIMO and cooperative-MISO are created in a distributed fashion. Over the two consecutive time phases of the cooperative-MIMO and cooperative-MISO schemes, the source terminal can potentially convey different information to the relay and destination terminals [5]. The cooperative-MIMO channel can be observed at the destination (D) terminal with the transmission sequence depicted in Figure 2.4. Such transmission scheme can realize an effective MIMO channel provided that the same MCS is used over the two phases [5]. This requires the two phases to have equal duration. The terms x_1 and x_2 represent different constellation points transmitted by the source (S) and the relay (R) terminals. The destination terminal sees a cooperative-MIMO channel with three non-zero channel coefficients. The cooperative-MISO and cooperative-SIMO channels can be observed at the destination terminal with the transmission sequences depicted in Figures 2.5 and 2.6, respectively. The destination terminal sees a cooperative channel with two non-zero channel coefficients. To observe a cooperative-MISO channel, the destination terminal does not receive the transmission of the source terminal during the first phase. In the second phase, the relay and

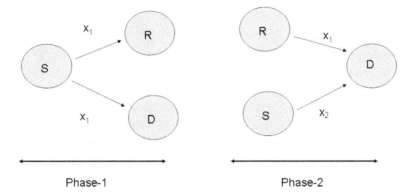

Figure 2.4 Creation of the cooperative-MIMO channel.

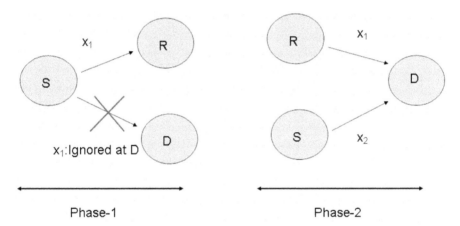

Figure 2.5 Creation of the cooperative-MISO channel.

the source transmit simultaneously. To achieve a cooperative-SIMO channel, the destination terminal receives during the both phases and the source terminal does not transmit during the second phase.

The relaying protocols analyzed in literature can be classified as [6]:

- Fixed protocols (fixed relaying): Relaying is always used in the second phase. The relays use the whole system frequency in the second phase.
- Adaptive protocols:
    (i) Whether the relay will be used or not is decided based on SNR conditions [27, 30]. This protocol is referred to as selection relaying in [30].

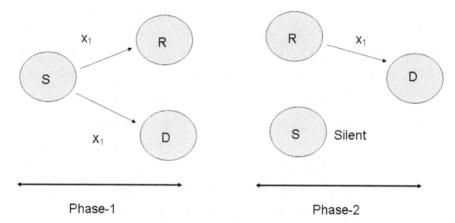

Figure 2.6 Creation of the cooperative-SIMO channel.

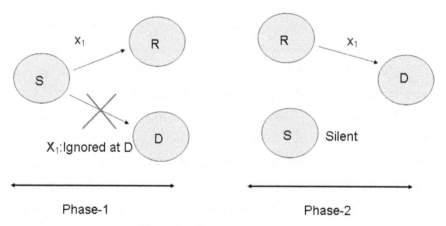

Figure 2.7 Conventional relaying

(ii) Simple-Adaptive Decode and Forward (AdDF): If the relay has decided to forward (upon successful decoding), the destination combines the signals received from the source and the relay. If the relay has decided not to forward (upon unsuccessful decoding), then it simply remains silent. In this case, the destination needs to rely on the samples stored in phase one. The possible silence if the relay has decided not to forward can be avoided by letting the source repeat its message which was transmitted in the first phase but could not be received correctly at the relay [30, 31].

(iii) Adaptive Decode and Re-encode (AdDR) [6]: The source and the relay perform distributed channel coding instead of repetition coding.

• Relaying with feedback from the destination: The relaying is used only when explicitly requested by the destination terminal.

2.2 Background on the Systems Considered

In this book, the IEEE 802.16e and IEEE 802.16j based wireless networks are considered. In this section, the background information on these networks is given.

2.2.1 The IEEE 802.16e Standard

This standard has been finalized. However, tremendous research activities pursue to further enhance its performance.

The Frame Structure: The frame structure of an IEEE 802.16e based cellular network is given in Figure 2.8 [3]. The frames have 5 ms of duration. As seen in the figure, the DL preamble is transmitted first. It is used for synchronization and channel estimation at the MSs. After scheduling and radio resource allocation which is done by the BS, the BS transmits in DL-Map (MAP) the control information on the index of the scheduled users on the slots, sub-channelization information, the used MCSs, etc. Hence, MAP messages serve as a map for the corresponding receivers. The preamble is followed by a Frame Control Header (FCH). It provides the frame configuration information such as the MAP message length, modulation and coding scheme and usable sub-channels, etc.

The data is transmitted over PHY bursts. In one burst, one or more than one user can be scheduled. One PHY burst contains single or multiple MAC-Protocol Data Units (PDUs). If Hybrid ARQ (HARQ) is not used, one MAC-PDU can only be transmitted within a single burst and shall not span over multiple bursts [32]. Each PHY burst consists of an integer number of slots. Each slot contains 48 data tones spanning over various OFDM symbols and sub-carrier frequencies. One tone represents one sub-carrier over one OFDM symbol. There are various predetermined slot configurations. For example, one slot may consist of 16 frequency diverse sub-carriers spanning over 3 contiguous OFDM symbols. The same MCS is used for transmission with a given burst. Fully Used Sub-Channelization (FUSC) and Partially Used Sub-

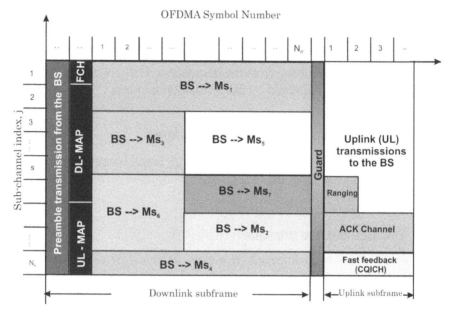

Figure 2.8 The frame structure of IEEE 802.16e standard [3].

channelization (PUSC) are diversity sub-carrier permutations. With these permutations, a sub-channel consists of non-contiguous sub-carriers distributed over the frequency band and the sub-carriers of each sub-channel are selected pseudo-randomly. This allows averaging the inter-cell interference over the frequency band and provides frequency diversity. The high mobility users (e.g., with speeds of 120 km/h) are scheduled on frequency diverse sub-carriers with PUSC or FUSC. The MCS to be used for each burst with PUSC and FUSC is selected based on long-term average channel statistics such as long-term average Signal-to-Interference-Plus-Noise-Ratio (SINR). The low mobility users are scheduled on sub-channels which consist of contiguous sub-carriers. Such sub-channelization is referred to as band-AMC and allows the use of opportunistic scheduling. With band-AMC sub-channelization, a bin consists of 8 data sub-carriers and one pilot sub-carrier. One slot consists of 48 data subcarriers with the following configurations. Nine contiguous sub-carriers (1 pilot + 8 data) spanning over 6 OFDM symbols, 18 contiguous (2 pilot + 16 data) sub-carriers spanning over 3 OFDM symbols, 27 contiguous (3 pilot + 24 data) sub-carriers spanning over 2 OFDM symbols, 54 contiguous (6 pilot + 48 data) sub-carriers spanning over 1 OFDM symbol.

Parameters	Values				
System bandwidth (MHz)	1.25	2.5	5	10	20
Sampling frequency (F_s, MHz)	1.429	2.857	5.714	11.429	22.857
Sample time ($1/F_s, nsec$)	700	350	175	88	44
FFT size (N_{FFT})	128	256	512	1024	2048
Subcarrier frequency spacing	11.16071429 kHz				
Useful symbol time ($T_b = 1/f$)	89.6 µs				
Guard time ($T_g = T_b/8$)	11.2 µs				
OFDMA symbol time ($T_s = T_b + T_g$)	100.8 µs				

Figure 2.9 The scalable OFDMA parameters of IEEE 802.16 standard [4].

With band-AMC sub-channelization, the AMC decisions are done per slot based on the instantaneous channel statistics such as instantaneous SINR.

The IEEE 802.16e standard provides scalable bandwidths and scalable number of sub-carriers. These modes are presented in Figure 2.9. As the system bandwidth and hence the total number of sub-carriers increases, the aggregate system throughput increases with an increasing transceiver complexity. Hence, for MSs with moderate transceiver capabilities, the scalable OFDMA mode with less number of sub-carriers can be used. Further details on the IEEE 802.16e standard can be found in [3, 32].

2.2.2 The IEEE 802.16j Standard

As of today, the IEEE 802.16j standard has not been finalized yet. The information presented in this section is based on [33]. Even if this standard is still a draft, it will be referred to as IEEE 802.16j standard in this book. The OFDM parameters[2] of the IEEE 802.16j standard is based on the IEEE

---

[2] Such as system bandwidth, sub-carrier spacing, total number of sub-carriers.

802.16e standard. However, the IEEE 802.16j standard enables multi-hop transmissions with the use of RSs. It defines two types of relays. These are transparent relay and non-transparent relay. The non-transparent relays can transmit preamble, DL and Uplink UL control information such as DL-MAP, UL-MAP, FCH, etc. It is seen as a BS to the end users. The transparent relays do not transmit control information. The users do not recognize the existence of the transparent relays. Transparent relaying can only be used with centralized scheduling done by the BS. Non-transparent relays enable distributed scheduling where they can decide on the radio resource allocation on the relay links. They can signal such allocation in their DL-MAP message. The relays inform the BS on their decoding status for each packet in the UL-ACKnowledgment ACK channel.

The IEEE 802.16j standard defines separate frame structures for transparent and non-transparent relays. For non-transparent relays, the standard supports more than two-hop transmissions and only one RS can be involved in reception of data destined to/from a given user. For transparent relays, only two hop transmissions are enabled and multiple RSs can be involved in reception and transmission of data destined to/from a given user. The frame structure contains both access zone and relay zone. In the relay zone, the RS is in either reception or transmission mode. In the DL-relay zone, $BS \rightarrow RS$ or $RS \rightarrow RS$ transmissions can be done. In the UL-relay zone, the $RS \rightarrow BS$ or $RS \rightarrow RS$ transmissions can be done. In the access zone, the MS is in either transmit or receive mode. In the DL-access zone, the $BS \rightarrow MS$, $RS \rightarrow MS$ or $BS \rightarrow RS$ transmissions can be done. In the UL-access zone, $MS \rightarrow BS$ or $MS \rightarrow RS$ transmissions are done. For more than two-hop transmissions with non-transparent relays, more than one relay zone and at least one access zone is used. On some of the sub-channels, the relays may choose to remain silent upon unsuccessful decoding of the packets transmitted by the BS in the DL-access zone. For non-transparent relaying, the IEEE 802.16j standard does not allow simultaneous transmissions from the BS and the non-transparent RS. Hence, the non-transparent relays use the entire frequency band when they are engaged in transmission. With transparent relaying in IEEE 802.16j standard, the BS may choose to transmit simultaneously with the RS in the DL transparent relay zone. However, the draft standard does not specify procedures on radio resource allocation for the RSs and the BS.

2.3 The System Model

In this section, the system model considered throughout this book is presented. Additional system considerations are detailed in each chapter.

DL transmissions in an OFDM(A)-TDD based two-hop cellular network are considered. A single cell is analyzed. Hence, the inter-cell interference is assumed to be zero. The analysis can be extended to multiple-cell scenario by considering the inter-cell interference. In a multi-cell environment, the SINR values should be used rather than SNR. If the interference is Gaussian distributed and independent than the noise, then in all the derivations, the SNR values can simply be replaced by the SINR values.

Infrastructure based relay terminal(s) deployed by a system operator to be exclusively used for relaying is considered. Only the closest RS is assigned to serve a given MS. This saves from the overhead and complexity in the system and still achieves tremendous gains in the performance. The source is a BS and the destinations are the MSs. The MS(s) are aware that the relays exist in the network. The developed algorithms do not increase the complexity of the MSs as compared to single-hop multi-antenna communication. A full buffer traffic where the BS has always data to transmit to a given user is assumed. The BS has equal data traffic for each user. Non-real time data traffic is assumed. Two main forwarding methods namely AF and DF have been considered. Repetition based relaying, where the relay repeats the information received from the BS is considered. All the terminals in the network have a single antenna which has omni-directional radiation pattern.

For all of the schemes, the average transmit energy of the source and relay terminals per sub–carrier is assumed to be constant and represented by E_S and E_R, respectively. This is necessary for multi-hop relay networks where the transmit power of the BS and RS are regulated independently to control the inter-cell interference in the system. Furthermore, in a practical system which uses AMC, power control does not bring significant throughput enhancement [34].

The throughput is defined as the number of bits per second per hertz and per channel use that are received correctly at the corresponding receiver. The goodput refers to the net throughput delivered to the upper layers. If relaying is used, the end-to-end throughput refers to the throughput delivered to the destination, i.e. a MS at the end of the two phases. If relaying is not used, then it simply refers to the throughput in a point-to-point link.

The design and evaluations are provided to evaluate and enhance the performance of IEEE 802.16j standard. As the w/o relay system, the IEEE

802.16e based cellular systems are considered for the performance comparisons [35].

OFDM adds a cyclic prefix to each symbol to mitigate inter symbol interference caused by multi-path propagation. If the cyclic prefix is longer than the delay spread of the wireless channel, then frequency selective fading caused by multi-path propagation can be converted into frequency flat fading at each sub-carrier. Adding a cyclic prefix to achieve this causes a small reduction in total transmission rate. Since this reduction is the same for each scheme, it is not taken into account in the throughput. The throughput is calculated based on the net end-to-end spectral efficiency with AMC.

Mobile users with relatively low speed are considered. For such users, the channel remains unchanged for the duration of a frame which consists of a certain number of OFDM symbols. Therefore, the link adaptations according to the obtained channel state information (e.g., instantaneous SINR) are feasible and effective. Hence, transmissions with band-AMC sub-channelization are considered [3]. A sub-channel is comprised of several contiguous sub-carriers with approximately equal channel coefficients. The FEC coding is applied on each sub-channel. This allows the code performance for each sub-channel to be modeled as being over an independent flat fading channel with its own SINR. The sub-channel index refers to the frequency dimension of the sub-channel. One burst with band-AMC subchannelization is comprised of one sub-channel. The bursts used for $BS \rightarrow MS$ or $RS \rightarrow MS$ transmissions are referred to as *access bursts*. Access bursts in the second hop have fixed duration and span the overall duration of the second hop. Each burst dedicated to $BS \rightarrow RS$ transmissions in the first hop is referred to as *relay burst*. The duration of the relay bursts depends on the relaying, forwarding and MCS chosen and it is adjusted such that the same number of bits is transmitted over each hop. Only one user is allocated to a given burst. However, depending on the channel conditions, one user can be allocated more than one bursts. It is assumed that the channel remain unchanged during one burst. For optimized operation with Band-AMC sub-channelization, the burst and/or sub-channel bandwidth should be less than the coherence bandwidth. Hence, the burst and sub-channel bandwidth is determined by the coherence bandwidth of the wireless channel. Via feedback, the BS has the ideal knowledge of the instantaneous SINR for each sub-channel and for each user regarding the $S \rightarrow R$, $S \rightarrow D$ and $R \rightarrow D$ links. This results in a CSI feedback overhead. Such overhead is present in systems which exploit link adaptation. Various CSI feedback algorithms can be used to reduce this overhead and are beyond the scope of this book. The design of link adaptation methods for high-mobility users requires further considerations, such as impact of imperfect channel state information, the form of channel state information to be used (e.g., [36]), etc.

It is assumed that each terminal (BS and RS) is transmitting to an MS at a given sub-carrier of an OFDM system in which each sub-carrier experiences frequency flat fading. Hence, each sub-carrier of the $S \rightarrow R$, $S \rightarrow D$ and $R \rightarrow D$ links are modelled as flat fading channels with given instantaneous SINR conditions. The instantaneous SINR in each link varies with variations in the channel. The instantaneous SINR and instantaneous SNR refers to the SINR and SNR in a given frame, respectively.

If DF based relaying is used, the MSs estimate the channel in the $BS \rightarrow MS$ and $RS \rightarrow MS$ links. If AF based relaying is used, the MSs estimate the channel in the $BS \rightarrow RS \rightarrow MS$ and $BS \rightarrow MS$ links. Each RS estimate the channel in the $BS \rightarrow RS$ link. Such estimations are done via the use of the pilot sub-carriers which are already available with the IEEE 802.16j and IEEE 802.16e standards. Imperfect channel estimation is not considered. It is assumed that no channel knowledge is available at the transmitters regarding the amplitude and phase of their transmission channels.

HARQ in the form of chase combining is used in order to combine the signals received by the BS and the RS. The considered cooperative diversity schemes are cooperative transmit diversity, cooperative receive diversity and cooperative selection diversity. ARQ is not considered. Analysis in this book can be extended to include ARQ.

2.3.1 Notation

The superscripts ^{H}, ^{T} and * stand for transpose conjugate, transposition and complex conjugation operations, respectively. Bold uppercase letters represent matrices and bold lower case letters represent vectors. $\mathbf{I}_m$ represents the $m \times m$ identity matrix. The term $\mathcal{E}$ denotes the expectation operation. The $\|.\|_F$ operator represents the Frobenius norm of its operand. $P(a)$ represents the probability of the event a. The term $*$ represents the convolution operation. The term $\mathcal{CN}(0, b)$ represents a zero mean circularly symmetric complex Gaussian random variable with mean 0 and variance b. The term $\mathbf{0}_{m,n}$ denotes a matrix of zeros with m rows and n columns. $H(\mathbf{a})$ represents the differential entropy of a given vector $\mathbf{a}$. $I(x; y)$ represents the mutual information of random variables x and y. $R_{\mathbf{xx}} \triangleq \mathcal{E}\{\mathbf{xx}^H\}$ is the autocorrelation function of random vector $\mathbf{x}$. $Q(x)$ is the Q-function which is defined as

$$Q(\sqrt{x}) = \frac{1}{\sqrt{2\pi}} \int_{\sqrt{x}}^{\infty} e^{-u^2/2} du.$$

2.3.2 Terminology

The subscript i, $i \in \{1, 2, \ldots, N\}$, represents the sub-carrier index where N represents the total number of sub-carriers in the OFDM system. The term j, $j \in \{1, 2, \ldots, J\}$, denotes the sub-channel index in the frequency domain. The total number of sub-channels is denoted by J. If sub-carrier i is within sub-channel j, then $\gamma_{SR,j} = \gamma_{SR,i}$, $\gamma_{SD,j} = \gamma_{SD,i}$ and $\gamma_{RD,j} = \gamma_{RD,i}$ represent the instantaneous SNR conditions at each sub-channel j. The index m where $m \in \{0, 1, 2, \ldots\}$ represents the OFDM symbol index in time domain.

In a point-to-point flat fading link with instantaneous SINR γ, let $\rho(\gamma) = R(\gamma)(1 - BLER(\gamma))$ represent the end-to-end throughput when AMC is used and the MCS which provides the highest throughput is selected. The term $BLER(\gamma)$ represents the block error rate with the selected MCS based on γ. The term $R(\gamma)$ in b/s/Hz represents the nominal rate of the selected MCS. If the selected coding rate is η and the selected M-ary modulation mode can provide a maximum rate of M b/s/Hz, then $R(\gamma) = M \times \eta$. For example if the selected MCS is 16-QAM with coding rate $\eta = 1/2$, then $R(\gamma) = 4/2$ b/s/Hz.

The term u, $u \in \{1, 2, \ldots, U\}$, denotes the MS index. The terms $\overline{\rho}^{\text{coopTxDiv1}}$, $\overline{\rho}^{\text{coopTxDiv2}}$, $\overline{\rho}^{\text{coopSDiv}}$ and $\overline{\rho}^{\text{direct}}$ represent the average end-to-end throughput per channel use that can be achieved with cooperative transmit diversity-1 and -2, cooperative selection diversity and w/o relay schemes, respectively. These terms represent the long-term average end-to-end throughput when relaying decisions are done dynamically.

2.3.3 Baseband Channel, Noise and Interference Models

The terms $h_{SR,i}$, $h_{SD,i}$ and $h_{RD,i}$ represent the frequency domain channel coefficient of sub-carrier i for $S \rightarrow R$, $S \rightarrow D$ and $R \rightarrow D$ links, respectively. These channel coefficients include the path loss and fast fading effects. Since a properly designed OFDM system converts frequency selective fading into frequency flat fading at each sub–carrier, it is assumed that each sub-carrier encounters frequency flat Rayleigh fading [37]. Hence, $|h_{SD,i}|$, $|h_{RD,i}|$ and $|h_{SR,i}|$ are modelled as Rayleigh flat fading random variables. The mobile(s) are scheduled on orthogonal radio resources. Hence, zero intra-cell interference is assumed. The interference is assumed to be caused by other cells in the network and modelled as a complex Gaussian random variable which is independent than the Additive White Gaussian Noise (AWGN). At a given sub-carrier i of the transmitted OFDMA symbol m, $n_{R,i}[m] \sim \mathcal{CN}(0, N_o^R)$ and $n_{D,i}[m] \sim \mathcal{CN}(0, N_o^D)$ represent the AWGN plus interference samples observed at the relay and destination terminals, respectively. The terms N_o^D and N_o^R represent the power spectral density of the receiver noise plus interference at each sub-carrier at the destination and relay terminals, respectively. With this model,

$$\gamma_{SR,i} = \frac{|h_{SR,i}|^2 E_S}{N_o^R}, \quad \gamma_{SD,i} = \frac{|h_{SD,i}|^2 E_S}{N_o^D} \quad \text{and} \quad \gamma_{RD,i} = \frac{|h_{RD,i}|^2 E_R}{N_o^D}$$

represent the instantaneous (i.e., short-term average obtained at a given DL frame) SINR conditions at sub-carrier i of $S \rightarrow R$, $S \rightarrow D$ and $R \rightarrow D$ links, respectively. The terms SNR_{SR}, SNR_{SD} and SNR_{RD} denote the long-term average (i.e., averaged over sufficiently many DL frames) SNRs in the $S \rightarrow R$, $S \rightarrow D$ and $R \rightarrow D$ links, respectively.

2.4 The End-to-End BER Performance of Cooperative Diversity in OFDM Based Wireless Relay Networks

In this section, the end-to-end BER performance of cooperative receive diversity with AF and DF based forwarding schemes has been analyzed comparatively. A hybrid forwarding scheme that adaptively decides to use either "AF" or "DF" or "no relay" according to the instantaneous SNR conditions between the source, relay and destination terminals has been proposed and analyzed. This decision is made based on analytically derived equations, which makes it possible to take into account all the link conditions in a relay network with low computational complexity. The BER performance is presented for cooperative receive diversity with AF, with DF, with the hybrid forwarding scheme and for non-cooperative (direct) transmission. The results show that the hybrid forwarding scheme significantly outperforms fixed relaying with AF or DF, and direct transmission. This gain comes especially from adaptively deciding on whether to use the relay or not. The results show that selection among AF and DF schemes does not bring significant performance gain when CSI is available at the source.

The use of either AF or DF at a relay terminal achieves different performance results under given Signal to Noise Ratio (SNR) conditions. The performance achieved with the DF scheme depends strongly on the ability of the RS to decode the received signal from the source terminal correctly. With AF, the noise at the RS is amplified and further propagated. The performance with AF depends strongly on the SNR of the source-to-relay ($S \rightarrow R$) link and amplification performed by the relay node. It might also be the case that relaying does not help at all. These observations motivate to consider a hybrid forwarding scheme which adaptively chooses the best transmission scheme according to the SNR conditions.

The hybrid forwarding scheme changes the type of the forwarding scheme used by a relay terminal on a per sub–carrier basis. In this section, the BER performance of the hybrid forwarding scheme has been analyzed and compared with that of fixed relaying with AF and DF, and also non-cooperative transmission.

Vajapeyam and Mitra [25] propose a hybrid space-time coding scheme for cooperative relaying. The study concentrates on frequency flat environments. Cooperative transmit diversity achieved with multiple relays is considered. Relays perform DF if instantaneous SNR condition in the $S \rightarrow R$ link is good. If not, relays use AF. However the performance of AF and DF schemes depends not only on the SNR condition of $S \rightarrow R$ link but also on the SNR conditions in the $S \rightarrow D$ and $R \rightarrow D$ links. Therefore, it is not enough to consider only the $S \rightarrow R$ link condition when deciding on the forwarding technique to be used by the relay terminal. Furthermore, hybrid forwarding mechanism with the consideration of frequency selective environments with OFDM networks has not been analyzed in the literature. Within frequency selective environments, sub-carriers of a properly designed OFDM network experience flat fading with different amplitudes. Therefore it would be beneficial for the relay terminals operating in OFDM networks to use the best forwarding scheme at each sub-carrier.

In this section, one source terminal, one destination terminal and one relay terminal assisting the transmission of the source terminal is considered. The transmission sequence which creates a cooperative-SIMO channel at the destination terminal has been considered. It is presented in Table 2.1. With this channel, cooperative receive diversity is extracted.

In order to make fair comparisons in BER analysis, the same end-to-end data rate has been used for different transmission schemes. For direct transmission scheme, Quadrature Phase Shift Keying (QPSK) modulation and $1/2$ rate convolutional coding have been used. With this scheme, the source terminal transmits on both phases. For cooperative relaying, uncoded transmission with QPSK modulation has been considered. This way, all the schemes transmit at a nominal end-to-end data rate of 1 b/s/Hz.

2.4.1 Input Output Relations in the First Phase

In the first phase, the source terminal transmits the constellation point $\sqrt{E_S}x_i[m]$ at time slot m on one of the sub-carriers, i.e. sub-carrier i. $\mathcal{E}[|x_i[m]|^2]$ is set to one. At time slot m, the relay and destination terminals receive the following baseband signals on sub-carrier i:

$$y_{R,i}[m] = h_{SR,i}\sqrt{E_S}x_i[m] + n_{R,i}[m] \tag{2.1}$$

$$y_{D,i}[m] = h_{SD,i}\sqrt{E_S}x_i[m] + n_{D,i}[m]. \tag{2.2}$$

2.4.2 Analysis of Forwarding Schemes for Relay Networks

In the following, the BER performance of cooperative receive diversity with AF and DF, and direct transmission scheme are analyzed and compared with each other. The derivations are presented for a given sub-carrier i.

2.4.2.1 Cooperative Receive Diversity with AF Scheme

In this section, the end-to-end BER achieved with cooperative receive diversity when the relay terminal uses the AF scheme is derived. The transmitted signal by the relay terminal using AF at sub-carrier i can be written as

$$s_R^{AF}[m+1] = \frac{\sqrt{E_R}}{\beta_i}y_{R,i}[m]. \tag{2.3}$$

The term

$$\alpha_i = \frac{\sqrt{E_R}}{\beta_i} = \frac{\sqrt{E_R}}{\sqrt{N_o^R(1+\gamma_{SR,i})}} \tag{2.4}$$

represents the amplification used at the RS at sub-carrier i determined such that $\mathcal{E}\{|s_{R,i}^{AF}|^2\} = E_R$ [28]. When the relay uses AF scheme, the received baseband signal at the destination terminal at sub-carrier i over the two phases can be written as:

$$
\begin{aligned}
\mathbf{y}^{AF} &= \begin{bmatrix} y_{D,i}[m] \\ y_{D,i}[m+1]^{AF} \end{bmatrix} \\
&= \begin{bmatrix} h_{SD,i}\sqrt{E_S}x_i[m] + n_D[m] \\ \frac{\sqrt{E_R}\sqrt{E_S}}{\beta_i}h_{SR,i}h_{RD,i}x_i[m] + \tilde{n}^{AF} \end{bmatrix} \\
&= \mathbf{h}^{AF}x_i[m] + \mathbf{n}^{AF},
\end{aligned}
\tag{2.5}
$$

where $\mathbf{h}^{AF}$ and $\mathbf{n}^{AF}$ are given by

$$\mathbf{h}^{AF} = \begin{bmatrix} \sqrt{E_S}h_{SD,i} \\ \frac{\sqrt{E_R}\sqrt{E_S}}{\beta_i}h_{SR,i}h_{RD,i} \end{bmatrix} \tag{2.6}$$

$$\mathbf{n}^{AF} = \begin{bmatrix} n_1 \\ n_2 \end{bmatrix} = \begin{bmatrix} n_D[m] \\ \frac{\sqrt{E_R}}{\beta_i}h_{RD,i}n_R[m] + n_D[m+1] \end{bmatrix}. \tag{2.7}$$

The term $\mathbf{h}^{AF}$ represents the cooperative-SIMO channel observed at the destination terminal after the two phases. The first row of $\mathbf{y}^{AF}$ represents the signal received at the destination terminal via the first phase. The second row of $\mathbf{y}^{AF}$ represents the signal received at the destination terminal via the second phase.

Assuming that the destination terminal has perfect knowledge of the channel coefficients, i.e., $h_{SR,i}$, $h_{SD,i}$ and $h_{RD,i}$, the received signals from the source and relay terminals are combined by MRC according to [10]:

$$z^{AF} = \mathbf{h}^{AF^H}\mathbf{y}^{AF}. \tag{2.8}$$

This is similar to chase combining in HARQ. The instantaneous SNR per symbol achieved after MRC combining in the AF mode can be derived from Equation (2.8) as follows:

$$\gamma_{s,i}^{AF} = \frac{\left(\gamma_{SD,i} + \frac{\gamma_{SR,i}\gamma_{RD,i}}{1+\gamma_{SR,i}}\right)^2}{\gamma_{SD,i} + \frac{\gamma_{SR,i}\gamma_{RD,i}}{1+\gamma_{SR,i}} + \frac{\gamma_{RD,i}^2\gamma_{SR}}{(1+\gamma_{SR,i})^2}}. \tag{2.9}$$

Let P_e^{AF} represent the BER at the destination terminal after MRC combining and ML detection at sub-carrier i. ML detection operation is equivalent to make a proper normalization, i.e., channel equalization, and then taking decisions. For a given flat fading condition at sub-carrier i, P_e^{AF} can be calculated in terms of $\gamma_{SR,i}$, $\gamma_{SD,i}$ and $\gamma_{RD,i}$ by

$$P_e^{AF} = Q\left(\sqrt{\gamma_{s,i}^{AF}}\right). \tag{2.10}$$

2.4.2.2 Cooperative Receive Diversity with DF Scheme

In this section, the end-to-end BER achieved with cooperative receive diversity when the relay terminal uses the DF scheme is derived. The relay uses

ML detection. Let $\hat{x}_i[m]$ with unit energy represent the detected constellation point at the relay . The performance of the DF scheme is severely affected by the detection errors at the relay terminal. If the relay terminal detects $x_i[m]$ wrong, then relaying causes error propagation. Let $P_e^{RS}(\gamma_{SR,i}) = P(\hat{x}_i[m] \neq x_i[m]) = Q(\sqrt{\gamma_{SR,i}})$ represent the probability of detection error at the RS. The probability of error propagation is therefore given by $P_e^{RS}(\gamma_{SR,i})$.

When the relay uses DF scheme, the received baseband signal at the destination terminal at sub-carrier i over two phases can be written as:

$$
\mathbf{y}^{DF} = \begin{bmatrix} y_{D,i}[m] \\ y_{D,i}[m+1]^{DF} \end{bmatrix}
$$
$$
= \begin{bmatrix} h_{SD,i}\sqrt{E_S}x_i[m] + n_D[m] \\ h_{RD,i}\sqrt{E_R}\hat{x}_i[m] + n_D[m+1] \end{bmatrix}. \tag{2.11}
$$

Let $\mathbf{h}^{DF}$ be defined as

$$
\mathbf{h}^{DF} = \begin{bmatrix} \sqrt{E_S}h_{SD,i} \\ \sqrt{E_R}h_{RD,i} \end{bmatrix}. \tag{2.12}
$$

When DF based forwarding is used, $\mathbf{h}^{DF}$ represents the cooperative-SIMO channel observed at the destination terminal at the end of the two phases.

Assuming that the destination terminal has perfect knowledge of $\mathbf{h}^{DF}$, the received signals from the source and relay terminals are combined by MRC as follows:

$$
z^{DF} = \mathbf{h}^{DF^H}\mathbf{y}^{DF}. \tag{2.13}
$$

For a given flat fading condition at sub-carrier i, let P_e^{DF} represent the BER at the destination terminal after MRC combining and ML detection. If $(\gamma_{SD,i} - \gamma_{RD,i}) \geqslant 0$, P_e^{DF} can be derived as

$$
P_e^{DF} = \left(1 - Q(\sqrt{\gamma_{SR,i}})\right)Q(\sqrt{\gamma_{SD,i} + \gamma_{RD,i}})
$$
$$
+ Q\left(\sqrt{\frac{(\gamma_{SD,i} - \gamma_{RD,i})^2}{\gamma_{SD,i} + \gamma_{RD,i}}}\right)Q(\sqrt{\gamma_{SR,i}}). \tag{2.14}
$$

If $(\gamma_{SD,i} - \gamma_{RD,i}) < 0$, P_e^{DF} can be derived as

$$
P_e^{DF} = \left(1 - Q(\sqrt{\gamma_{SR,i}})\right)Q(\sqrt{\gamma_{SD,i} + \gamma_{RD,i}})
$$
$$
+ \left(1 - Q\left(\sqrt{\frac{(\gamma_{SD,i} - \gamma_{RD,i})^2}{\gamma_{SD,i} + \gamma_{RD,i}}}\right)\right)Q(\sqrt{\gamma_{SR,i}}). \tag{2.15}
$$

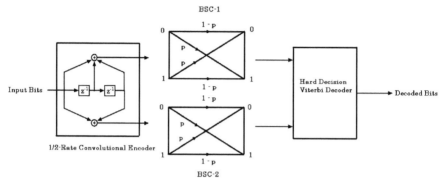

Figure 2.10 Encoder and decoder diagram for 1/2 convolutional code.

2.4.2.3 Direct Transmission Scheme

For a given flat fading condition at sub-carrier i, P_e^{Direct} represents the end-to-end BER with direct transmission scheme. In this section, P_e^{Direct} is derived. For the direct transmission scheme, the encoder and decoder diagram that is presented in Figure 2.10 is considered and hence P_e^{Direct} is upper-bounded by [38, 39]:

$$P_e^{\text{Direct}} < \sum_{d=5}^{\infty} \beta_d P_2(d), \tag{2.16}$$

where $\beta_d = 2^{d-5}(d - 4)$ and $P_2(d)$ is given by the following equations at $p = Q(\sqrt{\gamma_{SD,i}})$:

$$P_2(d) = \sum_{k=(d+1)/2}^{d} \binom{d}{k} p^k (1 - p)^{d-k} \tag{2.17}$$

for d odd and

$$P_2(d) = \sum_{k=d/2+1}^{d} \binom{d}{k} p^k (1 - p)^{d-k}$$
$$+ \frac{1}{2} \binom{d}{d/2} p^{d/2} (1 - p)^{d/2} \tag{2.18}$$

for d even [39, pp. 489–490]. Here, p represents the BER observed at the received coded bits. The term, β_d represents the distance properties of the convolutional encoder which is derived from its state diagram given in [38,

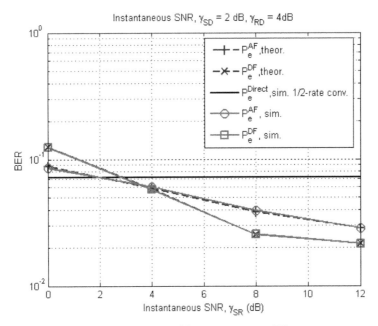

Figure 2.11 BER with the AF scheme (P_e^{AF}), DF scheme (P_e^{DF}) and direct transmission (P_e^{Direct}) versus $\gamma_{SR,i}$ with $\gamma_{SD,i} = 2$ dB and $\gamma_{RD,i} = 4$ dB.

pp.706]. d denotes the Hamming distance of the sequence of output bits corresponding to each branch from the sequence of output bits corresponding to the all-zero branch of the state diagram of the convolutional encoder. With the convolutional encoder considered in this section, the minimum hamming distance between any two codewords in the convolutional code is 5 and hence $d \geq 5$.

2.4.2.4 Performance Comparison of AF, DF and Direct Transmission Schemes

Figure 2.11 presents the theoretical and simulated BER curves for cooperative receive diversity and direct-transmission schemes for $\gamma_{SD,i} = 2$ dB and $\gamma_{RD,i} = 4$ dB. One can see from this figure that, if the SNR condition in the $S \rightarrow R$ link is relatively low, the AF outperforms the DF. On the other hand, if the SNR condition in the $S \rightarrow R$ link is good enough, the DF outperforms the AF. The important conclusion from the results in Figure 2.11 is that the best transmission scheme is different according to the SNR conditions. This motivates to consider a hybrid forwarding scheme which selects between AF, DF and direct transmission schemes at every sub-carrier.

2.4.3 Hybrid Forwarding Scheme

The proposed hybrid forwarding algorithm in this section is described as follows:

- The relay terminal measures $\gamma_{SR,i}$ in the DL slots. Via the help of channel reciprocity principle, it is assumed that the relay terminal obtains $\gamma_{RD,i}$ from the channel measurements of the $D \rightarrow R$ link. $\gamma_{SD,i}$ is assumed to be fed-back to the relay terminal by the destination terminal.
- With this information, the relay terminal calculates P_e^{AF}, P_e^{DF} and P_e^{Direct} simply with the derived equations
- At each sub-carrier, the relay terminal uses the scheme which provides the minimum BER. Therefore, for each sub-carrier, the destination terminal can obtain a BER equivalent to:

$$P_e^{HF} = \min\{P_e^{AF}, P_e^{DF}, P_e^{Direct}\}. \tag{2.19}$$

- The relay terminal informs the destination terminal on the type of forwarding scheme that is used at each sub-carrier. This information exchange needs to be done only when forwarding method is changed at each sub-carrier.
- The destination terminal performs appropriate MRC combining structure for each sub-carrier.

The transceiver structure of the relay terminal using the proposed hybrid forwarding scheme is presented in Figure 2.12. The proposed hybrid forwarding algorithm differs from already proposed forwarding schemes in the following aspects: (i) one OFDM symbol is transmitted by a relay terminal with different forwarding schemes, (ii) a relay terminal is able to take into account all the three link conditions when deciding on the best forwarding scheme for a given sub-carrier.

2.4.4 Numerical Results and Discussions

In this section, simulation results for average BER performance of hybrid forwarding (denoted by $\text{BER}^{HF} = \mathcal{E}\{P_e^{HF}\}$), fixed relaying with AF (denoted by $\text{BER}^{AF} = \mathcal{E}\{P_e^{AF}\}$) and DF (denoted by $\text{BER}^{DF} = \mathcal{E}\{P_e^{DF}\}$), and direct transmission (denoted by $\text{BER}^{Direct} = \mathcal{E}\{P_e^{direct}\}$) schemes are presented and compared with each other. Fixed relaying refers to the case where the relaying

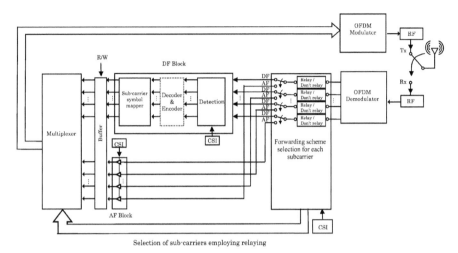

Figure 2.12 Transceiver block diagram of the relay terminal using hybrid forwarding scheme.

Table 2.2 System parameters.

Parameter	Value
System Bandwidth, B	20 MHz
Total number of data sub-carriers, N	400
Sub-carrier spacing, $\triangle f$	50 kHz
CP samples, N_g	100
$S \rightarrow R$ link: σ_{SR}, B/B_c, $B_c/\triangle f$	1.26 μsec, 125.6, 3
$S \rightarrow D$ link: σ_{SD}, B/B_c, $B_c/\triangle f$	0.23 μsec, 23.4, 17
$R \rightarrow D$ link: σ_{RD}, B/B_c, $B_c/\triangle f$	0.23 μsec, 23.4, 17

is always used on all the sub-carriers with the same forwarding scheme. In the analysis, the SNR conditions in the $S \rightarrow D$ and $R \rightarrow D$ links are fixed and SNR$_{SR}$ is varied.

Frequency selective, mutually independent channels are considered for $S \rightarrow R$, $S \rightarrow D$ and $R \rightarrow D$ links. A channel model for fixed wireless applications is used for the $S \rightarrow R$ link [40] and a fixed-to-mobile channel model is used for the $S \rightarrow D$ and $R \rightarrow D$ links [41]. The system parameters used in this section are outlined in Table 2.2. B_c represents the coherence bandwidth. σ_{SR}, σ_{SD} and σ_{RD} represent the rms delay spread in the $S \rightarrow R$, $S \rightarrow D$ and $R \rightarrow D$ links, respectively.

Figures 2.13 and 2.14 compare the average BER performance of the proposed hybrid forwarding scheme with that of fixed relaying with AF or DF, and direct transmission schemes. Figures 2.13 and 2.14 present the perform-

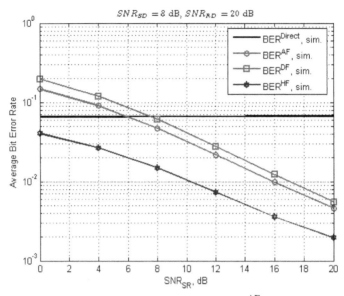

Figure 2.13 Average BER of conventional AF scheme (BER^{AF}), conventional DF scheme (BER^{DF}), direct transmission (BER^{Direct}) and proposed hybrid forwarding scheme (BER^{HF}) versus SNR$_{SR}$ with SNR$_{SD}$ = 8 dB and SNR$_{RD}$ = 20 dB.

ance results for SNR$_{SD}$ = 8 dB, SNR$_{RD}$ = 20 dB and SNR$_{SD}$ = 2 dB, SNR$_{RD}$ = 10 dB, respectively. As the results show, the proposed hybrid forwarding scheme achieves performance gains over direct transmission and fixed relaying and this gain becomes more obvious whenever SNR$_{RD}$ $\gg$ SNR$_{SD}$. For example as Figure 2.13 shows, the proposed hybrid forwarding scheme offers a gain of approximately 6 dB over fixed relaying with AF and 7 dB over fixed relaying with DF scheme at an average BER of 10^{-2}. When SNR$_{SR}$ $\gg$ SNR$_{RD}$ and SNR$_{SR}$ $\gg$ 1, then the BER performance at the destination with cooperative receive diversity depends on the SNR in the $R \rightarrow D$ and $S \rightarrow D$ links (see Equations (2.9), (2.14) and (2.15)). Since SNR$_{SD}$ and SNR$_{RD}$ are fixed in the figures presented in this section, the performance becomes independent of the SNR$_{SR}$ when SNR$_{SR}$ $\gg$ SNR$_{RD}$ and SNR$_{SR}$ $\gg$ 1. Therefore, after a certain average SNR threshold in the $S \rightarrow R$ link, the BER curves have a tail.

The results presented in this section show that, the selection among AF and DF schemes does not bring significant performance gain. However, the adaptive relaying in the form of to relay or not brings significant performance enhancement as compared to fixed relaying. As the results in this section

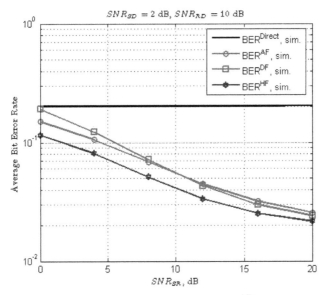

Figure 2.14 Average BER of conventional AF scheme (BER^{AF}), conventional DF scheme (BER^{DF}), direct transmission (BER^{Direct}) and proposed hybrid forwarding scheme (BER^{HF}) versus SNR$_{SR}$ with SNR$_{SD}$ = 2 dB and SNR$_{RD}$ = 10 dB.

show, link adaptive transmission mechanisms in wireless relay networks bring performance improvement. In the following chapters of this book, link adaptive transmission mechanisms which select the best forwarding, modulation, coding, cooperative diversity and the packet size have been developed to be used in wireless relay networks.

3

Link Adaptation and Selection Method for OFDM-Based Wireless Relay Networks*

3.1 Introduction

The error propagation problem inherent in DF-based relaying can be avoided by detecting the erroneously received packets via cyclic redundancy check or similar measures and forwarding only when the packets are correctly received by the relay. Such a scheme is referred to as simple-AdDF-based relaying [6, 31, 43]. With this scheme, the end-to-end throughput is limited not only by the post-processing SINR, but also the $S \rightarrow R$ link condition. For simple-AdDF, the selection of the MCS should take into account the SINR at the relay as the signal needs to be decoded correctly with high probability at the relay. However, for AF, the selection of the MCS can be based on the post-processing SINR at the destination, which might be larger than the SINR at the relay.

Erkip et al. analyzed adaptive modulation for coded cooperative systems achieved with DF-based relaying in [43]. The coded cooperation scheme considered in [43] uses the RS whenever it can correctly decode the transmitted packets. This means that, when the detection error at the RS is negligibly small, relaying is almost always used. However, if the SINR condition in the direct link allows to use a sufficiently high-rate MCS, then direct transmission might outperform relay-based transmission even if the $S \rightarrow R$ link condition is very good. This statement is proved later in this chapter. This is due to the

*The work presented in this chapter has been sponsored by Aalborg University and Telecommunication R&D Center-Samsung Electronics Co. Ltd., Suwon, Republic of Korea. It has been published in *Journal of Communications and Networks* (JCN), Special issue on "MIMO OFDM and Its Applications". This publication can be found in [42]. © 2007 JCN. Reprinted with permission from *Journal of Communications and Networks*, Special Issue on MIMO-OFDM and Its Applications, June 2007.

multiplexing loss inherent in wireless relaying. Hence, it is of utmost import- ance to consider the end-to-end throughput to decide on whether the relay should be used or not. The study in [27] proposed to make such a decision based on average channel conditions in the $S \rightarrow R$, $S \rightarrow D$ and $R \rightarrow D$ links. The design is made for flat fading channel conditions in the $S \rightarrow D$ and $S \rightarrow R$ links and a non-fading line of sight channel condition in the $R \rightarrow D$ link. For the design rules to decide whether to use the relay or not, the authors stated that due to high complexity simple decision rules cannot be developed to take into account small scale fading in all the links constituting a relay network. However, simple decision rules can be developed with the preparation of lookup tables which is provided in this chapter. Such lookup tables allow decisions for the selection of the most efficient MCS to be used.

In [30], Laneman et al. proposed a selection relaying scheme developed for frequency flat channels. The proposed scheme selects relaying only when the SNR in the $S \rightarrow R$ link is above a threshold which is determined solely by the channel capacity of the $S \rightarrow R$ link. When relaying is selected, either AF or DF-based relaying is used. The channel conditions in the $R \rightarrow D$ and $S \rightarrow D$ links are not considered in the proposed selection relaying in [30]. Even if the $S \rightarrow R$ link quality is above a threshold, relaying will not improve the system performance if the post-processing SINR cannot compensate for the multiplexing loss.

In summary, the afore-mentioned works have not taken into account all the instantaneous fading coefficients in a relay network and the end-to-end throughput performance at the same time. Consequently, these link adaptation and selection mechanisms cannot guarantee that the end-to-end throughput performance is not worse than that of w/o relay and fixed relaying.

In this chapter, an end-to-end link adaptation and selection method for OFDM-based wireless relay networks is proposed. Relaying is selected (with the best forwarding scheme) only when it can improve the end-to-end throughput as compared to that of w/o relay transmissions. This selection is made based on the instantaneous SINR conditions of the links constituting the relay network. The proposed link adaptation and selection method dy- namically selects the best transmission method at each sub-channel. Simple and efficient rules for

1. end-to-end link adaptation with AMC
2. link selection (transmission with or w/o relay)

are provided based on lookup tables. A frame structure to enable the proposed link adaptation and selection method in an OFDMA-TDD-based cellular wireless relay network has been provided.

The structure of this chapter is as follows. Section 3.2 describes the system model. In Section 3.3, the end-to-end throughput performance of the simple-AdDF, AF and w/o relay schemes are presented and compared to each other. The proposal for an end-to-end link adaptation and selection method for wireless relay networks is presented in Section 3.4 and its superior performance has been shown. Conclusions and future works are drawn in Section 3.5.

3.2 System Model

The analysis is based on the following system model.

The forwarding schemes considered in this chapter are AF and simple-AdDF. More complex forwarding schemes other than AF and simple-AdDF can be included into the framework of link adaptation and selection for wireless relay networks. A single relay is considered. The analysis in this chapter can be extended to the multi-relay scenarios. The transceiver structure of the relay terminal using AF and DF-based relaying is presented in Chapter 2. During the down-link sub-frame no feedback is provided by the MS since it is engaged in data reception. In the first phase of a given relaying scheme, the RS using simple-AdDF-based relaying performs the following operations. It demodulates the OFDM symbols received from the BS. After the FFT operation, the RS decodes the signal at each sub-channel and performs cyclic redundancy check to determine the blocks that are correctly received. The RS re-encodes and buffers only the blocks that are correctly received over a given sub-channel. Before the second phase starts, the RS broadcasts control information on which sub-channels it could detect the blocks correctly and on which sub-channels it could not. This control signalling will only be needed for simple-AdDF-based relaying. In Section 3.4, it is shown that this overhead is negligibly small in a practical system setting. In the second phase, the RS and the BS make cooperative transmission only for the blocks which are correctly received at the RS. This study can be extended to the case where in the second phase, the BS repeats the blocks that are erroneously received by the RS.

Ideal synchronization is assumed. With simple-AdDF-based relaying, the MSs do not need to know any channel state information regarding the $S \rightarrow R$ link condition. However, with the AF-based relaying, the channel state

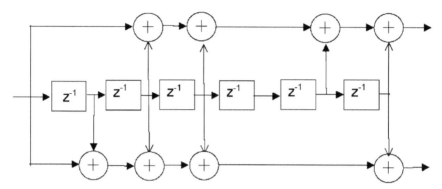

Figure 3.1 The convolutional encoder with code rate 1/2, constraint length 7 ($K = 7$) and code generator [171, 133] (in octal).

information regarding the $S \to R$ link is necessary at the MSs which in turn increases the complexity.

The MAC-PDUs are transmitted in FEC blocks. Each block includes a CRC. A block is discarded if at least one bit is in error. The receivers use cyclic redundancy check to detect block errors where it is assumed that the undetected block error probability is negligibly small [44]. An FEC block is comprised of 96 coded bits [45]. The block error probability (i.e., block error rate) is evaluated for the reception of an FEC block. The modulation modes that are considered in this chapter are: Binary Phase Shift Keying (BPSK), QPSK, 16-Quadrature Amplitude Modulation (QAM) and 64-QAM. The FEC is considered in the form of convolutional coding with code rates: "1(no-coding), 1/2, 2/3, 3/4, 5/6, 7/8" [3]. Each combination of the modulation and coding modes gives one AMC mode, i.e., MCS. The convolutional encoder that is used in this chapter and in Chapter 4 is given in Figure 3.1. This encoder has a code rate of 1/2. The other code rates are generated via puncturing.

3.2.1 The Considered Relaying Schemes

In this chapter, various relaying schemes such as cooperative-MISO, cooperative-SIMO and cooperative-MIMO are considered to achieve cooperative diversity. Conventional relaying scheme is considered as well. These relaying schemes were introduced in Chapter 2. An efficient end-to-end link adaptation method for each of these relaying schemes is presented.

Table 3.1 Transmission sequence to achieve cooperative diversity over a cooperative-MIMO channel with Alamouti scheme.

First phase	First phase	Second phase	Second phase	...
m-th symbol	$m+1$-st symbol	$m+2$-nd symbol	$m+3$-rd symbol	...
$S \to R, D$ $(x_{1,i})$	$S \to R, D$ $(-x_{2,i}^*)$	$S \to D\,(x_{2,i})$, $R \to D\,(x_{1,i})$	$S \to D\,(x_{1,i}^*)$, $R \to D\,(-x_{2,i}^*)$	...

3.2.1.1 Cooperative Transmit Diversity-1

This scheme refers to the cooperative transmit diversity achieved with cooperative-MIMO scheme. To achieve cooperative transmit diversity, cooperative-space-time coding is used by the BS and the RS terminals as given in Table 3.1. In the table, $x_{1,i}$, $x_{2,i}$ denote the transmitted constellation points at sub-carrier i of the corresponding link. To achieve cooperative transmit diversity-1 scheme, the two symbols $x_{1,i}$ and $x_{2,i}$ are transmitted over four OFDM symbol intervals using the Alamouti scheme. The symbol indices m and $m+1$ represent the transmitted OFDM symbols during the first phase. The symbol indices $m+2$ and $m+3$ represent the transmitted OFDM symbols during the second phase. To achieve diversity, the destination terminal processes the resulting four outputs received during the two phases.

For AF and simple-AdDF-based relaying, the cooperative-MIMO channel observed at the MS at the end of the two phases is given by

$$\mathbf{H}^{AF} = \begin{bmatrix} \sqrt{E_S}h_{SD,i} & 0 \\ \frac{\sqrt{E_R E_S}}{\beta_i}h_{SR,i}h_{RD,i} & \sqrt{E_S}h_{SD,i} \end{bmatrix} \tag{3.1}$$

and

$$\mathbf{H}^{DF} = \begin{bmatrix} h_{1,1} & h_{1,2} \\ h_{2,1} & h_{2,2} \end{bmatrix} = \begin{bmatrix} \sqrt{E_S}h_{SD,i} & 0 \\ \sqrt{E_R}h_{RD,i} & \sqrt{E_S}h_{SD,i} \end{bmatrix}, \tag{3.2}$$

respectively [5]. The term β_i is given by Equation (2.4).

At the end of the two phases, the MS performs space-time decoding of the signals received from the BS and the RS terminals during the two phases. The MS achieves further SINR gain by combining the space time decoded symbols with the corresponding symbols received in the first phase. Finally, with AF-based relaying, the post-processing SINR achieved at the MS is given by

$$\gamma_{s,i}^{AF} = \frac{\left(2\gamma_{SD,i} + \frac{\gamma_{SR,i}\gamma_{RD,i}}{(1+\gamma_{SR,i})}\right)^2}{\gamma_{SD,i} + \left(1 + \frac{\gamma_{RD,i}}{(1+\gamma_{SR,i})}\right)\left(\frac{\gamma_{SR,i}\gamma_{RD,i}}{(1+\gamma_{SR,i})} + \gamma_{SD,i}\right)}. \tag{3.3}$$

With AF-based relaying, the MS can achieve an end-to-end throughput given by $0.5\rho(\gamma_{s,i}^{AF})$. The factor of 0.5 accounts for the fact that, two time phases with equal duration is needed with AF-based relaying.

At the end of the two phases with simple-AdDF-based relaying, the MS creates the following receive vectors to perform space-time decoding:

$$\mathbf{y}_1 = \begin{bmatrix} y_{D,i}[m] \\ y_{D,i}[m+2] \end{bmatrix} = \mathbf{H}^{DF} \begin{bmatrix} x_{1,i} \\ x_{2,i} \end{bmatrix} + \begin{bmatrix} n_1 \\ n_2 \end{bmatrix},$$

$$\mathbf{y}_2 = \begin{bmatrix} y_{D,i}[m+1] \\ y_{D,i}[m+3] \end{bmatrix} = \mathbf{H}^{DF} \begin{bmatrix} -x_{2,i}^* \\ x_{1,i}^* \end{bmatrix} + \begin{bmatrix} n_3 \\ n_4 \end{bmatrix}, \tag{3.4}$$

where the 2×2 cooperative-MIMO channel, i.e., $\mathbf{H}^{DF}$, is given by Equation (3.2). This effective cooperative-MIMO channel can be achieved at the cost of a multiplexing loss of $1/2$ for each modulation mode. For the transmitted OFDM symbol m, $y_{D,i}[m]$ denotes the received baseband signal at sub-carrier i at the destination. The MS forms a signal vector $\mathbf{y}$ according to

$$\mathbf{y} = \begin{bmatrix} \mathbf{y}_1 \\ \mathbf{y}_2^* \end{bmatrix} = \begin{bmatrix} h_{1,1} & h_{1,2} \\ h_{2,1} & h_{2,2} \\ h_{1,2}^* & -h_{1,1}^* \\ h_{2,2}^* & -h_{2,1}^* \end{bmatrix} \begin{bmatrix} x_{1,i} \\ x_{2,i} \end{bmatrix} + \begin{bmatrix} n_1 \\ n_2 \\ n_3^* \\ n_4^* \end{bmatrix} \tag{3.5}$$

$$= \mathbf{H}_{\text{eff}}\mathbf{x} + \mathbf{n}, \tag{3.6}$$

where $\mathbf{x} = [x_{1,i} \quad x_{2,i}]^T$ and

$$\mathbf{n} = [n_1 \quad n_2 \quad n_3^* \quad n_4^*]^T$$

$$= \left[n_{D,i}[m] \quad n_{D,i}[m+2] \quad n_{D,i}^*[m+1] \quad n_{D,i}^*[m+3] \right]^T. \tag{3.7}$$

The MS now performs MRC for $\mathbf{y}$ according to

$$\mathbf{z} = \mathbf{H}_{\text{eff}}^H \mathbf{y}. \tag{3.8}$$

Assuming that the RS decodes the transmitted symbols by the BS correctly, $\mathbf{z}$ becomes

$$\mathbf{z} = \|\mathbf{H}^{DF}\|_F^2 \mathbf{x} + \tilde{\mathbf{n}} \tag{3.9}$$

where $\tilde{\mathbf{n}} = \begin{bmatrix} \tilde{n}_1 \\ \tilde{n}_2 \end{bmatrix} = \mathbf{H}_{\text{eff}}^H \mathbf{n}$ and $E[\tilde{\mathbf{n}}] = \mathbf{0}_{2,1}$. The resulting effective noise power can be calculated as

$$E[|\tilde{n}_1|^2] = E[|\tilde{n}_2|^2] = N_0^D \|\mathbf{H}^{DF}\|^2. \tag{3.10}$$

Assuming that the RS decodes the transmitted symbols by the BS correctly, the post-processing instantaneous SNR at each sub-carrier i achieved after MRC at the MS can be derived from Equations (3.2), (3.9) and (3.10) as

$$\gamma_{s,i}^{DF} = \frac{(\|\mathbf{H}^{DF}\|_F^2)^2}{E[|\tilde{n}_1|^2]} = \frac{\|\mathbf{H}^{DF}\|_F^2}{N_o^D}$$

$$= 2\gamma_{SD,i} + \gamma_{RD,i}. \tag{3.11}$$

Hence, second order diversity can be achieved for each symbol transmitted by the BS.

The MCS to be used in both phases is selected based on $\rho = \min\{\rho(\gamma_{s,i}^{DF}), \rho(\gamma_{SR,i})\}$. This is analogous to $\min(\gamma_{s,i}^{DF}, \gamma_{SR,i})$. If $\rho = \rho(\gamma_{SR,i})$ then the MCS is decided based on $\gamma_{SR,i}$, otherwise, the MCS is selected based on $\gamma_{s,i}^{DF}$. This way, the system can keep negligible error rates at the RS if $\gamma_{SR,i} > \gamma_{s,i}^{DF}$ and at the MS if $\gamma_{s,i}^{DF} > \gamma_{SR,i}$. Furthermore, the system can keep acceptable error rates at the RS if $\gamma_{SR,i} < \gamma_{s,i}^{DF}$. With this MCS selection method, the MS can achieve an end-to-end throughput given by

$$\rho_{\text{coop-div}}^{\text{AdDF}} = \begin{cases} 0.5\rho(\gamma_{s,i}^{DF})P_c(\gamma_{SR,i}), & \text{if } \rho = \rho(\gamma_{s,i}^{DF}), \\ 0.5\rho(\gamma_{SR,i})P_c(\gamma_{s,i}^{DF}), & \text{if } \rho = \rho(\gamma_{SR,i}). \end{cases}$$

$P_c(\gamma)$ represents the probability of correct reception of a block with the selected MCS (based on ρ) over a flat fading channel with SINR γ.

3.2.1.2 Cooperative Transmit Diversity-2

This scheme refers to the cooperative transmit diversity achieved with the cooperative-MISO scheme. To achieve diversity, cooperative space time coding can be used by the BS and the RS terminals [5] and Alamouti space time coding is used [11]. This scheme achieved with DF-based relaying does not necessitate two phases with equal duration since the MS does not exploit any

Table 3.2 Transmission sequence to achieve diversity over a cooperative-MISO channel with Alamouti scheme.

First phase m-th symbol	First phase $m + 1$-st symbol	Second phase $m + 2$-nd symbol	Second phase $m + 3$-rd symbol	...
$S \rightarrow R$ ($x_{1,i}$)	$S \rightarrow R$ ($-x_{2,i}^*$)	$S \rightarrow D\ (x_{2,i})$, $R \rightarrow D\ (x_{1,i})$	$S \rightarrow D\ (x_{1,i}^*)$, $R \rightarrow D\ (-x_{2,i}^*)$	...

signal transmitted in the first phase. Hence, the MCS for each phase is chosen independently. This leads to an efficient use of radio resources. For DF-based relaying, the MCS per hop is selected based on the post-processing SINR at the corresponding receiver node, i.e., either the RS or the MS. For AF-based relaying, the decoding is done only at the MS at the end of two phases. Hence the MCS is chosen according to the post-processing SINR observed at the MS. The time divisioned transmission structure to achieve cooperative transmit diversity-2 scheme is presented in Table 3.2. The two symbols $x_{1,i}$ and $x_{2,i}$ are transmitted via sub-carrier i over four OFDM symbol intervals using the Alamouti scheme. The first phase consists of the first two OFDM symbols, i.e, symbols m and $m+1$. The second phase consists of the third and fourth OFDM symbols, i.e., symbols $m + 2$ and $m + 3$. The destination terminal processes the resulting two outputs received during the second phase. In the following sections (1.A and 1.B), the input output relations with the cooperative transmit diversity-2 is presented for both AF and simple-AdDF-based relaying. The transmission sequence presented in Table 3.2 has been considered.

1.A Input Output Relations with AF-Based Relaying
At the end of the two phases at a given sub-carrier i, the MS creates the following received signal vector for each transmitted symbol:

$$
\mathbf{y} = \begin{bmatrix} y_{D,i}[m + 2] \\ y_{D,i}^*[m + 3] \end{bmatrix} = \begin{bmatrix} h_1 & h_2 \\ h_2^* & -h_1^* \end{bmatrix} \begin{bmatrix} x_{1,i} \\ x_{2,i} \end{bmatrix} + \begin{bmatrix} n_1 \\ n_2^* \end{bmatrix}
$$
$$
= \mathbf{H}_{\text{eff}}\mathbf{x} + \mathbf{n}, \tag{3.12}
$$

where $\mathbf{x} = [x_{1,i}\ \ x_{2,i}]^T$ and $\mathbf{n} = [n_1\ \ n_2^*]^T$. At the end of the two phases, the 1×2 cooperative-MISO channel achieved with AF-based relaying is given by

$$\mathbf{h}^{AF} = \begin{bmatrix} h_1 & h_2 \end{bmatrix}$$

$$= \begin{bmatrix} \frac{\sqrt{E_R E_S}}{\beta_i} h_{SR,i} h_{RD,i} & \sqrt{E_S} h_{SD,i} \end{bmatrix}, \tag{3.13}$$

where β_i is given by Equation (2.4). The noise components n_j where $j \in \{1, 2\}$ are given by $n_j = n_{D,i}[m + j + 1] + h_{RD,i} \frac{\sqrt{E_R}}{\beta_i} n_{R,i}[m + j - 1]$. The MS performs maximum ratio combining on $\mathbf{y}$ according to

$$\mathbf{z} = \mathbf{H}_{\text{eff}}^{H} \mathbf{y} = \|\mathbf{h}^{AF}\|_F^2 \mathbf{x} + \tilde{\mathbf{n}}, \tag{3.14}$$

where $\tilde{\mathbf{n}} = \mathbf{H}_{\text{eff}}^{H} \mathbf{n} = [\tilde{n}_1 \ \tilde{n}_2]^T$. At a given sub-carrier i, the post-processing instantaneous SINR achieved after maximum ratio combining at the MS can be derived from Equations (3.13), (3.14) as

$$\gamma_{s,i}^{AF} = \frac{\left(\|\mathbf{h}^{AF}\|_F^2\right)^2}{\mathcal{E}\{|\tilde{n}_1|^2\}} = \frac{\gamma_{SD,i} + \frac{\gamma_{SR,i}\gamma_{RD,i}}{(1+\gamma_{SR,i})}}{1 + \frac{\gamma_{RD,i}}{(1+\gamma_{SR,i})}}. \tag{3.15}$$

After combining at a given sub-carrier, the MS observes an effective point-to-point link with a post-processing SINR as given by Equation (3.15).

With AF-based relaying at a given sub-carrier i, the end-to-end throughput achieved with an Alamouti-scheme-based cooperative transmit diversity is finally given by

$$\rho_{tx-\text{div}}^{AF} = 0.5\rho(\gamma_{s,i}^{AF}), \tag{3.16}$$

where the MCS for both of the transmission phases is selected based on $\gamma_{s,i}^{AF}$.

1.B *Input Output Relations with Simple-AdDF-Based Relaying*
The cooperative-MISO channel achieved with simple-AdDF-based relaying is given by $\mathbf{h}^{DF} = \begin{bmatrix} h_1 & h_2 \end{bmatrix} = \begin{bmatrix} \sqrt{E_R} h_{RD,i} & \sqrt{E_S} h_{SD,i} \end{bmatrix}$. Since the RS relays only when it can decode the blocks correctly, the post-processing instantaneous SINR achieved at the MS after an Alamouti-scheme-based space-time decoding can be derived as [10, 11]:

$$\gamma_{s,i}^{DF} = \frac{\|\mathbf{h}^{DF}\|_F^2}{N_o^D} = \gamma_{SD,i} + \gamma_{RD,i}. \tag{3.17}$$

With simple-AdDF-based relaying at a given sub-carrier i, the end-to-end throughput is finally given by

$$\rho_{tx-\text{div}}^{AdDF} = \frac{\rho(\gamma_{SR,i})\rho(\gamma_{s,i}^{DF})}{R(\gamma_{SR,i}) + R(\gamma_{s,i}^{DF})}, \tag{3.18}$$

where the MCS in the first phase is selected based on $\gamma_{SR,i}$ and the MCS in the second phase is selected based on $\gamma_{s,i}^{DF}$ in order to optimize the end-to-end throughput.

3.2.1.3 Cooperative Receive Diversity

This scheme was analyzed in detail in Chapter 2 and it achieves diversity via a cooperative-SIMO channel. The cooperative-SIMO channel achieved with AF-based relaying is given by Equation (2.6). To achieve cooperative receive diversity via maximum ratio combining, two phases with equal duration is needed since both the BS and the RS need to transmit with the same MCS to constitute a cooperative-SIMO channel [5]. To achieve receive diversity, the MS combines the signals received from the BS and the RS via maximum ratio combining. Over the sub-carrier where the cooperative receive diversity with AF-based relaying is used, the post-processing instantaneous SINR obtained at the MS after maximum ratio combining is given by Equation (2.9) [28]. The simple-AdDF-based relaying achieves the post-processing instantaneous SINR, i.e., $\gamma_{s,i}^{DF}$, as given in Equation (3.17). With AF-based relaying at a given sub-carrier i, the end-to-end throughput achieved by the cooperative receive diversity is given by

$$\rho_{rx-div}^{AF} = 0.5\rho(\gamma_{s,i}^{AF}), \tag{3.19}$$

where the MCS per transmission phase is selected based on $\gamma_{s,i}^{AF}$. For simple-AdDF-based relaying, the proposed end-to-end link adaptation method chooses the MCS to be used by the BS and the RS based on

$$\rho = \min\{\rho(\gamma_{s,i}^{DF}), \rho(\gamma_{SR,i})\}. \tag{3.20}$$

With this MCS selection, the end-to-end throughput, i.e., $\rho_{coop-div}^{AdDF}$, is given by Equation (3.12).

3.2.1.4 Conventional Relaying

In this book, conventional relaying uses rate adaptive relaying. Therefore, the end-to-end throughput with conventional relaying corresponds to the end-to-end throughput of cooperative transmit diversity-2 with $\gamma_{SD,i} = 0$. If the BS dynamically selects the best scheme among conventional relaying and w/o relay scheme, cooperative selection diversity can be provided to the MS. The cooperative selection diversity is introduced and analyzed in Chapter 4.

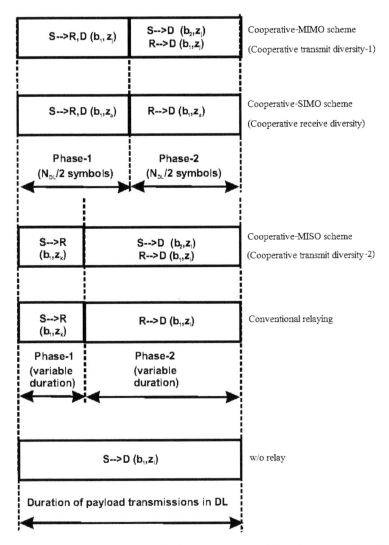

Figure 3.2 The time-divisioned transmission sequence to achieve the cooperative diversity schemes with DF.

3.2.1.5 Summary of the Transmission Sequence of Various Cooperative Diversity Schemes

In summary, the time-divisioned transmission sequences to achieve the cooperative diversity schemes presented in this section are visualized in Figure 3.2 for DF-based relaying. N_{DL} denotes the total number of OFDM symbols including the first and second phases. The terms z_i, z_j, z_k, z_l and z_a denote the MCS that is used in each phase. The terms $\mathbf{b_1}$ and $\mathbf{b_2}$ represent the different vector of bits transmitted by the source and relay terminals in each phase. The cooperative transmit diversity-1 and cooperative receive diversity use fixed rate relaying [46]. The distinguishing feature of the cooperative transmit diversity-2 and conventional relaying with DF is the fact that they allow the use of rate adaptive relaying [46] and hence the duration of the first and second phase of these schemes are optimized independently.

3.3 End-to-End Throughput Performance Comparison of AF, Simple-AdDF and w/o Relay Schemes

3.3.1 The Lookup Table

For each SINR value γ, the lookup table stores

1. The index of the MCS which provides the highest throughput,
2. $\rho(\gamma)$, b/s/Hz,
3. $P_c(\gamma)$ for each and every MCS available.

Such a lookup table is needed and used for the proposed link adaptation and selection method which will be described in detail in the next section. The lookup table is created for discrete values of γ with resolution 0.1 dB. Figure 3.3 presents the stored $\rho(\gamma)$ for each discrete value of γ. The value of $\rho(\gamma)$ depends on the block length. This study can be extended to accommodate FEC blocks with different lengths.

3.3.2 The End-to-End Throughput Performance

In this section, the end-to-end throughput performance of the simple-AdDF, AF and w/o relay schemes is compared to each other. The results are presented for each of the relaying schemes described in the previous section while using the proposed AMC decision rules. In the following, the results are plotted versus $\gamma_{SR,i}$ for given $\gamma_{SD,i}$ and $\gamma_{RD,i}$.

(1) *Cooperative transmit diversity-2*: Figure 3.4 presents simulation results with $\gamma_{SD,i} = 20$ dB and $\gamma_{RD,i} = 6$ dB. The results in this figure show that, even if the $S \to R$ link has a very good SINR level, relaying may not improve the end-to-end throughput. Hence, the relaying should be selected only if it can improve the end-to-end throughput. Furthermore, in the figure, the simple-AdDF scheme is throughput limited as compared to the AF scheme. However, over this region, relaying does not improve the end-to-end

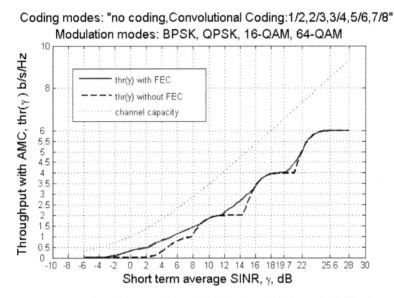

Figure 3.3 Representation of the stored $\rho(\gamma)$ in the lookup table for each SINR value γ. The γ values are discrete with resolution 0.1 dB.

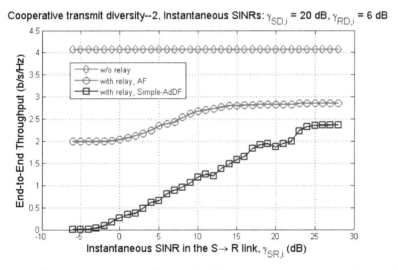

Figure 3.4 Cooperative transmit diversity-2: The instantaneous end-to-end throughput achieved with AF, simple-AdDF and w/o relay schemes versus $\gamma_{SR,i}$ with $\gamma_{SD,i} = 20$ dB and $\gamma_{RD,i} = 6$ dB.

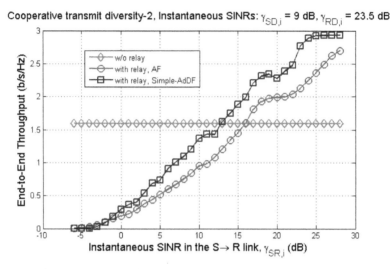

Figure 3.5 Cooperative transmit diversity-2: The instantaneous end-to-end throughput achieved with AF, simple-AdDF and w/o relay schemes versus $\gamma_{SR,i}$ with $\gamma_{SD,i} = 9$ dB and $\gamma_{RD,i} = 23.5$ dB.

performance. The main conclusions for the relative performance of AF and simple-AdDF-based relaying should be done over the region where relaying improves the performance. Figure 3.5 presents the simulation results with $\gamma_{SD,i} = 9$ dB and $\gamma_{RD,i} = 23.5$ dB. The results show that relaying can improve the end-to-end throughput for higher $\gamma_{SR,i}$ and the simple-AdDF-based relaying can provide significant throughput gain as compared to AF-based relaying. For the cooperative transmit diversity-2, over the region where relaying improves the end-to-end throughput, the simulation results indicate that the AF-based relaying cannot outperform the simple-AdDF-based relaying for all the instantaneous SINR conditions in the $S \rightarrow R, S \rightarrow D$ and $R \rightarrow D$ links. This is due to the fact that, with simple-AdDF-based relaying, the link adaptation can adjust the transmission rates in each phase independently and hence can use the radio resources more efficiently than the AF scheme. Note that the end-to-end throughput with simple-AdDF-based relaying is not monotonically increasing as the nominal transmission rates take discrete values (see Equation (3.18)).

(2) *Cooperative receive diversity*: Over the instantaneous SINR region where relaying improves the end-to-end throughput, the simulation results indicate that the AF-based relaying can outperform both the end-to-end

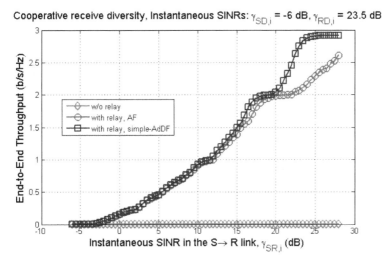

Figure 3.6 Cooperative receive diversity: The instantaneous end-to-end throughput achieved with AF, simple-AdDF and w/o relay schemes versus $\gamma_{SR,i}$ with $\gamma_{SD,i} = -6$ dB and $\gamma_{RD,i} = 23.5$ dB.

throughput achieved with simple-AdDF-based relaying and w/o relay scheme as much as 0.02 b/s/Hz. Hence, the AF scheme cannot outperform simple-AdDF-based relaying significantly over this SINR region. Figure 3.6 presents the simulation results with $\gamma_{SD,i} = -6$ dB and $\gamma_{RD,i} = 23.5$ dB. The results in this figure show that simple-AdDF-based relaying has a throughput gain of as much as 0.72 b/s/Hz compared to AF based relaying.

(3) *Cooperative transmit diversity-1*: Over the instantaneous SINR region where relaying improves the end-to-end throughput, the simulation results with cooperative-transmit diversity-1 indicate that the AF-based relaying cannot provide significant throughput enhancement over simple-AdDF-based relaying. Figure 3.7 presents the simulation results with $\gamma_{SD,i} = 1.5$ dB and $\gamma_{RD,i} = 23.5$ dB. The results show that simple-AdDF-based relaying can provide significant throughput enhancement to both AF-based relaying (as much as 0.73 b/s/Hz) and w/o relay (as much as 2.48 b/s/Hz) transmissions.

(4) *Conventional relaying*: For all the combinations of $\gamma_{SR,i}$ and $\gamma_{RD,i}$, the simulation results with conventional relaying indicate that, AF-based relaying cannot outperform simple-AdDF-based relaying. Simple-AdDF-based relaying on the other hand can provide a throughput gain of as much as 0.71 b/s/Hz compared to AF-based relaying.

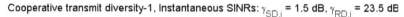

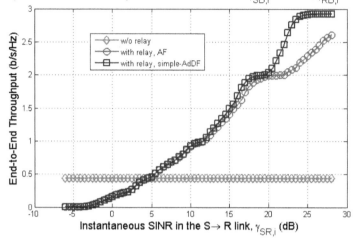

Figure 3.7 Cooperative transmit diversity-1: The instantaneous end-to-end throughput achieved with AF, simple-AdDF and w/o relay schemes versus $\gamma_{SR,i}$ with $\gamma_{SD,i} = 1.5$ dB and $\gamma_{RD,i} = 23.5$ dB.

Consequently, the results presented in this section motivate a link adaptation and selection method that dynamically selects the best scheme among "simple-AdDF and w/o relay" based on the CSI.

3.4 The Proposed End-to-End Link Adaptation & Selection Method and Its Performance Evaluation

3.4.1 The Proposed End-to-End Link Adaptation & Selection Method

With the observations presented in the previous section, a link adaptation and selection method for OFDM/OFDMA-TDD-based cellular wireless relay networks is proposed. This proposal is characterized by the following. The link adaptation and selection is done by the BS (in order to have centralized control). The channel state information regarding the SINRs in each sub-channel of $S \rightarrow R$, $S \rightarrow D$ and $R \rightarrow D$ links are obtained at the BS (i.e., fed-back by the users) at the end of each UL sub-frame via the Channel Quality Indication CHannel (CQICH) [3]. Let this channel state information be referred to as $\gamma_{SR,j}$, $\gamma_{RD,j}$ and $\gamma_{SD,j}$ for each sub-channel j. The BS then calculates the post-processing SINR for each sub-channel (i.e., $\gamma_{s,j}^{DF}$, $\gamma_{s,j}^{AF}$) and for each scheme based on the derived equations in Section 3.2.1. The scheme (among AF, simple-AdDF and w/o relay) which provides the highest end-to-end throughput is then selected for each sub-channel based on $\gamma_{SR,j}$, $\gamma_{RD,j}$ and $\gamma_{SD,j}$. This selection is done with the proposed AMC decisions and it is based on a lookup table described in the previous section. With the lookup table, the throughput for a given instantaneous SINR γ is determined by reading the corresponding throughput. For AF-based relaying, γ is determined by the post-processing SINRs derived in Section 3.2.1. For simple-AdDF-based relaying, $\rho(\gamma_{s,j}^{DF})$, $\rho(\gamma_{SR,j})$, $P_c(\gamma_{s,j}^{DF})$ and $P_c(\gamma_{SR,j})$ are read from the lookup table (refer to Equation (3.12)). Then, the end-to-end throughput for each scheme is calculated based on the equations derived in Section 3.2.1. Finally the transmission scheme and the MCS providing the highest end-to-end throughput is determined for each sub-channel and for each transmission phase, i.e., hop.

3.4.2 The Frame Structure

In order to use the proposed end-to-end link adaptation and selection method in a practical system, the frame structure should be designed carefully. In Figure 3.8, a frame structure is presented as a possible solution to enable the proposed link adaptation and selection in an OFDM(A)-TDD-based cellular wireless relay network where the users are within the coverage area of the BS and some of them are in the coverage area of both the BS and the RS. This frame structure takes into account of unreliable $BS \rightarrow RS$ link conditions by providing a control information channel to the relays. Over this channel, the RS can send control information on which sub-channels it could detect the transmissions correctly. In Figure 3.8, a single RS is considered. Before data transmissions start, the BS selects the best scheme among with or w/o relay schemes for each user and for each sub-channel. Then, the BS schedules the users with efficient scheduling algorithms developed for wireless relay networks. Such scheduling algorithms are developed in Chapter 4. After scheduling, the BS broadcasts in DL-MAP which user is scheduled on which sub-channel and for each sub-channel which scheme (among AF, simple-AdDF and w/o relay) and MCS are used. After the first phase and a guard interval for the transmit/receive turnaround time, the RS starts the transmission. The RS first transmits its preamble and control information.[1] The BS needs to receive this information as well and hence it has to switch from the receive mode to the transmit mode before data transmissions start. This

---

[1] The BS and RS can also transmit their preambles simultaneously.

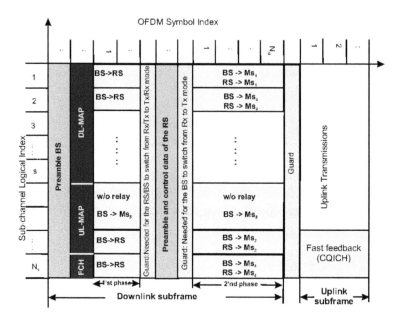

Figure 3.8 The frame structure as a possible solution to enable the proposed link adaptation and selection in an OFDMA-TDD-based cellular wireless relay network where the users are within the coverage of the BS and the RS.

necessitates a guard interval. Finally, transmissions with the selected schemes start in the second phase. In Figure 3.8, cooperative transmit diversity-2-based transmissions are considered over the sub-channels where relaying is selected by the BS. In the first and second phase, data transmission at each sub-channel take place which is designated for a given MS. For cooperative diversity schemes which do not necessitate two phases with equal duration, the duration of the first phase can be adjusted based on the average SINR in the $S \rightarrow R$ link.

For cooperative selection diversity, the frame structure developed for a two-hop cellular network with infrastructure-based relays is presented in Figure 3.9. This frame structure is developed for reliable, LOS $BS \rightarrow RS$ links, where the relays are deployed at strategic positions in the cell. Based on this model, it is assumed that the block error rate at the relays are negligible. Multiple relays in the cell are considered. Only the closest RS serves to a given user. All the relays and users listen to the DL-MAP information. In the figure, the duration of the second phase is fixed. If conventional relaying is selected for a given user, then the duration of the first phase at each sub-

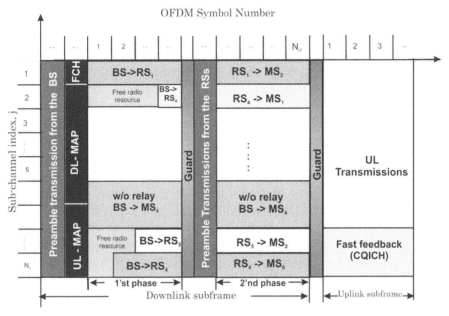

Figure 3.9 Frame structure for low mobility users in two-hop cellular networks with infrastructure-based relays. CQICH stands for the channel quality indicator channel provided in the IEEE 802.16e standard. $\{RS_1, RS_2, \ldots\}$ denote the different relays. $\{MS_1, MS_2, \ldots\}$ denote the different scheduled users at each sub-channel.

channel can be variable depending on the MCS chosen for the second phase. Since the $BS \rightarrow RS$ links have good channel conditions, the duration of the first phase at each sub-channel can be shorter than or equal to that of second phase. This allows having some free radio resources in the first phase as seen in Figure 3.9. These free radio resources can be used by the scheduler to schedule additional $S \rightarrow D$ (w/o relay) transmissions. For conventional relaying or cooperative transmit diversity-2 schemes, the scheduler has the freedom to optimize the locations of the free radio resources and the radio resources to be used for the $S \rightarrow R$ transmissions. In scheduling with such optimization, the priority should be given to the $S \rightarrow D$ transmissions as fading will be more severe as compared to LOS $S \rightarrow R$ links. The guard interval in the DL sub-frame is needed in order for the RS to switch from the receive mode to the transmit mode.

3.4.3 Performance Evaluation

In the following, the simulation setup is introduced for the performance evaluation of the proposed end-to-end link adaptation and selection method. The system parameters of the IEEE 802.16e standard with a total of 2048 sub-carriers with a system bandwidth of 22.4 MHz is considered [3]. In the simulations, only 360 point FFT is done in order to prevent very long simulation time. The system operates at a carrier frequency of 2.5 GHz. The frames have 5 ms duration. 44 out of 48 OFDM symbols in a given frame are reserved for data transmission where 24 of them are reserved for down-link transmissions [3]. The frame structure depicted in Figure 3.8 is considered. The down-link sub-frame is divided into two phases where each phase can have duration of 12-OFDM symbols. A multi-path wireless channel model is considered for the $S \rightarrow D$ and $R \rightarrow D$ links with an rms delay spread of 0.231 µs [41]. A sub-carrier spacing of 22.4 MHz/2048 $= 10.94$ kHz is assumed. This corresponds to a 90% coherence bandwidth of 8 sub-carriers and a 50% coherence bandwidth of 80 sub-carriers in the $S \rightarrow D$ and $R \rightarrow D$ links [47]. For the $S \rightarrow R$ link, a wireless channel model that is developed for fixed wireless applications in [40, 48] is used. For this link, an rms delay spread of 0.264 µs and a LOS K factor of 1 is considered. This corresponds to a 50% coherence bandwidth of 70 sub-carriers for a sub-carrier spacing of 10.94 kHz. The $S \rightarrow D$ and $R \rightarrow D$ links are assumed to be non-line-of-sight, i.e., with a K factor of zero. One sub-channel consists of nine consecutive sub-carriers (eight for data, one for pilot) over m consecutive OFDM symbols where $m, m \in \{2, 3, 6, 12\}$, depends on the selected modulation mode [3]. A single user with a speed of up to 7.7 km/h is considered such that the 50% coherence time is equal to 10 ms [47]. A total of 40 sub-channels is simulated and all of them have been allocated to a single user. At a carrier frequency of 2.5 GHz, a speed of 7.7 km/h results in a maximum Doppler frequency shift of 17.8 Hz [47]. Since this maximum Doppler shift is much smaller than the sub-carrier spacing (i.e., 0.2% of the sub-carrier spacing), the receiver at the MS does not see significant performance degradation due to inter-carrier-interference caused by Doppler frequency shift [49]. These system parameters enables to assume a block fading channel which remains flat within a given sub-channel in a frame.

At the end of the first phase, the RS using the simple-AdDF-based relaying sends control information on its decoding status for each block received from the BS. The overhead for this signalling will be smaller than or equal to

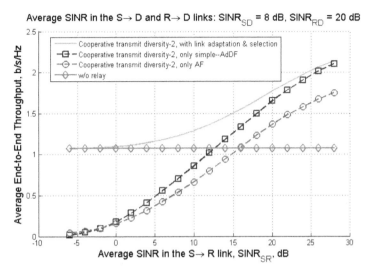

Figure 3.10 Cooperative transmit diversity-2: The average end-to-end throughput achieved with the proposed link adaptation method as compared to non-adaptive schemes where one type of transmission method is used. The average SINRs: $\text{SINR}_{SD} = 8$ dB and $\text{SINR}_{RD} = 20$ dB.

$N_{\text{sub-ch}} \times 3$ bits per block/(5 ms),[2] which equals to 24 kb/s or 11×10^{-4} b/s/Hz in a 22.4 MHz system bandwidth with a number of sub-channels, i.e., $N_{\text{sub-ch}}$, equal to 40. Hence, this signalling will not cause significant overhead. This overhead can further be reduced if the RS sends information on only the blocks that are not correctly received (efficient when $S \rightarrow R$ link is reliable) or that are correctly received (efficient if the $S \rightarrow R$ link is not reliable).

In the following, the simulation results are provided for the performance evaluation of the proposed link adaptation and selection method. Figures 3.10–3.13 present:

1. the average end-to-end throughput achieved with the proposed link adaptation and selection method,
2. the average end-to-end throughput achieved with (i) simple-AdDF, (ii) AF and (iii) w/o relay schemes which are used over all the sub-channels.

---

[2] Since the MCSs considered in this chapter can provide a maximum transmission rate of 6 b/s/Hz, the system can send at most $12/(96/(8 \times 6)) = 6$ blocks per nine contiguous sub-carriers in the first phase which necessitates three bits per block for identification of each block.

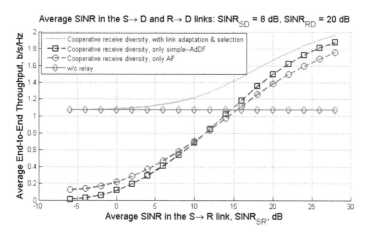

Figure 3.11 Cooperative receive diversity: The average end-to-end throughput achieved with the proposed link adaptation method as compared to non-adaptive schemes where one type of transmission method is used. The average SINRs: $SINR_{SD} = 8$ dB and $SINR_{RD} = 20$ dB.

For the cooperative transmit diversity-2 scheme, the results in Figure 3.10 are obtained versus the average SINR condition in the $S \rightarrow R$ link where the average SINR conditions in the $S \rightarrow D$ and $R \rightarrow D$ links are fixed to 8 and 20 dB, respectively. The results presented in Figure 3.10 show that, the average end-to-end throughput performance obtained via the proposed link adaptation and selection method is always better than or equal to that of (i) w/o relay transmissions and (ii) transmissions where relaying is used over all the sub-channels.

For cooperative receive diversity, the results in Figure 3.11 are obtained versus the average SINR condition in the $S \rightarrow R$ link where the average SINR conditions in the $S \rightarrow D$ and $R \rightarrow D$ links are fixed as 8 and 20 dB, respectively. The results show that the transmissions with the proposed link adaptation and selection method improves the performance and the AF-based relaying cannot outperform simple-AdDF-based relaying over the region where relaying improves the throughput.

For cooperative transmit diversity-1 scheme, the results in Figure 3.12 are obtained versus the average SINR condition in the $S \rightarrow R$ link where the average SINR conditions in the $S \rightarrow D$ and $R \rightarrow D$ links are fixed as 8 and 20 dB, respectively. The same conclusion as in cooperative receive diversity is drawn.

For conventional relaying scheme, the results in Figure 3.13 are obtained versus the average SINR condition in the $S \rightarrow R$ link where

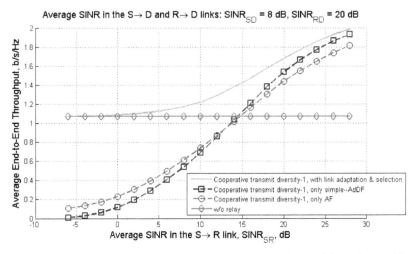

Figure 3.12 Cooperative transmit diversity-1: The average end-to-end throughput achieved with the proposed link adaptation method as compared to non-adaptive schemes where one type of transmission method is used. The average SINRs: $SINR_{SD} = 8$ dB and $SINR_{RD} = 20$ dB.

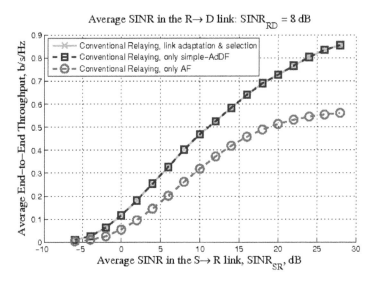

Figure 3.13 Conventional relaying scheme: The average end-to-end throughput achieved with the proposed link adaptation method as compared to non-adaptive schemes where one type of transmission method is used. The average SINR: $SINR_{RD} = 8$ dB.

the average SINR condition in the $R \rightarrow D$ link is fixed as 8 dB. The results in this figure show that the AF scheme is throughput limited as compared to the simple-AdDF scheme. Hence, the proposed link adaptation and selection method when used with conventional relaying selects the simple-AdDF-based relaying.

3.5 Conclusions and Future Work

In this chapter, an end-to-end link adaptation and selection method for wireless relay networks has been proposed. A frame structure in order to enable transmissions with this proposal is developed for OFDM(A)-TDD-based cellular wireless relay networks. Simple and efficient AMC decision rules have been developed for wireless relay networks. Such rules are able to take into account the fading conditions in all the wireless links constituting a relay network whereby the end-to-end throughput is optimized. The investigations show that, the end-to-end throughput performance with the proposed link adaptation and selection method is always better than or equal to that of (i) w/o relay transmissions and (ii) relayed transmissions where relaying is always used. On the other hand, the investigations show that, in a practical system setting, the AF-based relaying cannot outperform simple-AdDF-based relaying over the region where relaying improves the end-to-end throughput as compared to w/o relay transmissions. Hence, transmissions with the DF scheme is promising as the error propagation can be avoided by error detection techniques which are already inherent in wireless transmissions. The DF-based relaying provides superior performance with properly designed AMC. Future work out of this chapter includes:

1. The effect of synchronization on OFDM(A)-based relay networks using the proposed link adaptation and selection method.
2. The design of the proposed link adaptation and selection method with imperfect or reduced channel state information.
3. The design of the proposed link adaptation and selection method for users with high speeds. For such users, the sub-channel allocation is done on frequency diverse sub-carriers to provide frequency diversity [3]. In this case, new lookup tables should be prepared with average SINR conditions as instantaneous SINR conditions cannot be taken into account due to high variations in the channel.
4. The evaluation of the reduction in power consumption at the MS by introducing the cooperative diversity only when necessary.

5. The investigations from channel capacity point of view where the performance with AF and simple-AdDF-based relaying are compared over the region where relaying improves the end-to-end performance. This investigation is presented in Section 4.8 and the conclusions drawn form such analysis agree with the conclusions drawn in this chapter.

6. The design of a fully adaptive system which dynamically selects not only the best forwarding scheme but also the best relaying scheme. This design is presented in Chapter 4.

Acknowledgments

The author would like to thank Euntaek Lim, Persefoni Kyritsi, Maciej Portalski, Hugo Simon Lebreton, Simone Frattasi, M. Imadur Rahman, the reviewers of the publication in [42] and the other involved researchers working at the "R&D Center, Samsung Electronics Co. Ltd." and the Center for TeleInFrastruktur (CTIF) for their valuable comments on the work presented in this chapter.

4

Cooperative Diversity Schemes and User Scheduling with Fixed Relays for IEEE 802.16j*

4.1 Introduction

For given channel conditions, the end-to-end throughput of various coope-rative diversity schemes is different from each other. Hence, in order to optimize the end-to-end throughput, the radio resource allocation can not only use the best MCS but also choose the best scheme that is providing the highest end-to-end throughput. In this chapter, scheduling and radio resource alloc-ation that can be used for the IEEE 802.16j-based wireless relay networks are developed. The considered cooperative diversity schemes are cooperative transmit diversity, cooperative receive diversity and cooperative selection di-versity. A fully adaptive cooperative diversity scheme which selects the best transmission scheme among various cooperative diversity schemes and the w/o relay scheme is developed and analyzed. For the emerging IEEE 802.16j-based wireless networks, this chapter provides a performance evaluation of various cooperative diversity schemes in a multi-user environment. The in-vestigations are provided comparatively. The beneficial cooperative diversity scheme in terms of spectral efficiency and complexity has been identified for low mobility users.

The rest of this chapter is organized as follows. In the following section, prior arts are presented. In Section 4.3, the system model is described. In Section 4.4.1, the properly designed various cooperative diversity schemes are presented. In Section 4.4.2, various scheduling algorithms for two-hop

*The work presented in this chapter has been sponsored by Aalborg University, Tele-communication R&D Center-Samsung Electronics Co. Ltd., Suwon, Republic of Korea and Carleton University, Canada. A patent on this work has been filed by the Korean Intellectual Patent Office and can be found in [50]. A part of this work has been published in [51]. © 2008 IEEE. Reprinted with permission from *Proceedings of the IEEE Wireless Communications and Networking Conference (WCNC)*, April 2008.

cellular networks are developed. The deployment of the relays at strategic positions in the cell has been analyzed in Section 4.5. Based on the schedulers developed in Section 4.4.2, the performance of various cooperative diversity schemes has been analyzed in the context of the emerging IEEE 802.16j standard. Such analysis is given in Sections 4.6 and 4.7. In Section 4.8, the performance of various relaying schemes is compared from an information theoretic perspective. Conclusions and future works are drawn in Section 4.9.

4.2 Prior Arts

The literature for scheduling of the users in a wireless relay network considers the design either from information theoretic point of view or based on only the SNR or SINR conditions which do not consider the multiplexing loss inherent in relaying (e.g., [52–55]). The emerging IEEE 802.16j standard may allow w/o relay transmissions in the second phase. However, the current standard does not specify how the radio resource allocation will be done [33]. In [56, 57], Kaneko and Popovski divide the frame into two phases and allocate the whole frequency band to the relays in the second phase. Such predefined frame structure can reduce the CSI overhead [57]. However, it causes throughput loss as compared to the scheme which uses the best scheme with or w/o relay for each sub-channel [42, 57]. When a user's SNR condition for a given sub-channel remains unchanged during the whole frame (e.g., for low mobility users), the optimal scheme for a given sub-channel will not change. For example, if such a user is scheduled w/o relay in the first phase, that user should be scheduled w/o relay in the second phase as well. For optimal operation, the schedulers developed in this chapter may choose to use the w/o relay transmissions in the second phase.

The radio resource allocation and link adaptation mechanisms developed for wireless relay networks so far have either focused on only one relaying scheme or did not make performance comparison among various relaying schemes in a multi-user scenario.

4.3 System Model

With the infrastructure-based relay terminals deployed in the multi-hop cellular networks, the $BS \rightarrow RS$ ($S \rightarrow R$) links can be very robust since the relays are fixed and can be deployed at strategic positions in the cell. The RS can have a significant receive antenna gain and height. These factors make it practical for the relay to decode the signals it receives from the BS with negligible error. This is verified in Section 4.5. Hence, it is assumed that the $S \rightarrow R$ links are reliable, i.e., the relay can decode the signals received from the source with negligible error. As the benefits of simple-AdDF-based forwarding is shown in Chapter 3, the relays use simple-AdDF.

In order to obtain optimization for transmissions with either cooperative transmit diversity-2 or conventional relaying, a flexible frame structure has been assumed where the duration of the first and second phases for each sub-channel can be adjusted based on the nature of the cooperative diversity scheme. For e.g., the frame structure presented in Figure 3.9 is considered for cooperative selection diversity. For cooperative transmit diversity-1 and cooperative receive diversity schemes, the free radio resources in the first phase (presented in Figure 3.9) do not exist. This is due to the fact that these schemes necessitate fixed rate relaying [46] and hence cannot extract the benefit of good $S \rightarrow R$ links to shorten the duration of the first phase. For each cooperative diversity scheme analyzed in this chapter, the BS schedules the users with the best scheme among the schemes with or w/o relay. If w/o relay scheme is chosen for a given sub-channel, it is used in both phases.

Multiple users are considered. Since CSI is available at the BS, AMC is used for each sub-channel and for each frame based on the end-to-end link adaptation and selection method developed in Chapter 3 [42]. Due to its low complexity thanks to the use of a lookup table and the derived equations for the end-to-end throughput, this method makes it possible to do the user scheduling (i.e., prioritization of the users for radio resource allocation) together with path selection for each sub-channel. The considered MCSs are the following. The modulation modes are BPSK, QPSK, 16-QAM and 64-QAM. The considered FEC is convolutional coding with the following code rates: 1/2, 2/3, 3/4, 5/6, 7/8 and 1 [3].

The MAC-PDUs are transmitted in FEC blocks. Each FEC block includes a CRC. If at least one bit in a block is received in error, that block is assumed to be in error and hence discarded. To detect block errors, the receivers use cyclic redundancy check. It is assumed that, with cyclic redundancy check, the probability of an undetected block error is negligibly small [44].

Since the BS and the RS are geographically separated, the signals simultaneously transmitted by them arrive at an MS with a time offset. This time offset depends on the distance of the MS to the RS and the BS. As shown in Section 6.1, this time offset problem does not cause significant ISI if the relays add an appropriate delay to their transmission. Hence, it is assumed that the ISI caused by the time offset is negligible. Thanks to the robust

$S \rightarrow R$ links, the RS can estimate its local oscillator mismatch with that of BS precisely and can compensate for this offset before transmission. This way, the MSs experience only one carrier frequency offset which can be easily compensated at the MS. Hence, it is assumed that the carrier frequency offset does not cause significant throughput degradation.

The average end-to-end throughput is calculated per channel use, i.e., the average is taken over the radio resources allocated to the users. The coverage radius r is defined as the radius of the coverage area where the user throughput is greater than or equal to 0.5 b/s/Hz with probability greater than or equal to p. The outage probability is defined as $P_{out} = 1 - p = 1 - P(\rho \geq 0.5)$. The term ρ represents the instantaneous end-to-end throughput achieved at a given MS. The MS is assumed to be in outage when $P_{out} > 0.05$. The BS is at the center of the cell. The positions of the relays are marked with "x".

4.3.1 Simulation Setup

In this section, the simulation setup considered in this chapter is introduced.

For all the schemes, the DL frames have 5 ms of duration [3]. Multiple users with speeds up to 7.7 km/h have been assumed such that the 50% coherence time is greater than or equal to 10 ms [47]. Hence, link adaptation and scheduling will be effective for each frame. The duration of the second-phase is fixed to 12 OFDM symbols. The first phase can use up to 12 OFDM symbols. One sub-channel is comprised of eight data sub-carriers and one pilot subcarrier. An FEC block is comprised of 96 coded bits. The block error probability is evaluated for the reception of an FEC block. The transmission of one FEC block requires t consecutive OFDM symbols. The term t, $t \in \{2, 3, 6, 12\}$ depends on the selected MCS with AMC. The system parameters of the scalable OFDMA mode with 2048 sub-carriers with a system bandwidth of 22.4 MHz is considered [3]. In the simulations, only 540 point FFT is done in order to prevent very long simulation time. This corresponds to the simulation of a total of 60 sub-channels.

A path-loss exponent of 3 and a Rician K factor of 10 are used for the $S \rightarrow R$ links. For these links, a wireless channel model developed in [40, 48] is used. The selected model has a 90% coherence bandwidth of 17 sub-carriers. For the $R \rightarrow D$ and $S \rightarrow D$ links, a path-loss exponent of 3.5 is used. For these links, a NLOS wireless channel model developed for mobile applications is used [41]. The selected model has a 90% coherence bandwidth of eight sub-carriers. The above-mentioned channel models and the mobility of the users enable to assume a block fading channel which is frequency flat within a given burst. A total of 540 sub-carriers include $540/8 = 67.5$ coherence bandwidths, therefore, the results will not change if a larger fractions of sub-carriers were simulated.

The MSs are assumed to have a height of 1.5 m. The BS and RS heights are 32 and 10 m, respectively. A carrier frequency of 2.5 GHz is considered. Based on these assumptions, path-loss at each link is calculated accordingly [40]. In order to evaluate the performance versus distance dependent path loss, the effect of shadowing is not considered. A total of 60 users have been simulated to create the multi-user environment.

The total Effective Isotropic Radiated Power (EIRP) from the BS is fixed as 57.3 dBm [3]. It is assumed that the transmit power from each relay station is 3 dB less than that of BS [58]. Since the relay terminals are simpler than a BS, it is assumed that the transmit antenna gain from each relay is 7 dBi less than that of BS. Hence, the total EIRP from each relay station is fixed as 47.3 dBm.

4.4 Cooperative Diversity Schemes and Scheduling of the Users in a Two-Hop Cellular Network

4.4.1 Cooperative Diversity Schemes

In the following, the cooperative diversity schemes analyzed in this chapter are introduced. For all the schemes, the transmission for each user in each phase occurs at a given sub-channel j. Cooperative transmit diversity-1, cooperative transmit diversity-2 and cooperative receive diversity schemes were introduced in Chapter 3. In addition to these schemes, cooperative selection diversity and adaptive cooperative diversity schemes are introduced in the context of infrastructure-based relay terminals.

4.4.1.1 Cooperative Transmit Diversity-1

As derived in Chapter 3, the post-processing SINR achieved at the MS with cooperative transmit diversity-1 is given by $\gamma_{\text{post},j}^{\text{coopTxDiv1}} = 2\gamma_{SD,j} + \gamma_{RD,j}$. Since it is assumed in this chapter that the $S \rightarrow R$ link can support the highest rate MCS with negligible decoding error, the MCS for a given sub-channel j is chosen based on $\gamma_{\text{post},j}^{\text{coopTxDiv1}}$. With such link adaptation at a sub-channel j, the end-to-end throughput is given by

$$\rho_j^{\text{coopTxDiv1}} = 0.5\rho(\gamma_{\text{post},j}^{\text{coopTxDiv1}}). \tag{4.1}$$

4.4.1.2 Cooperative Transmit Diversity-2

For this scheme, the MCS to be used in the first phase is chosen based on $\gamma_{SR,j}$ for each sub-channel j. For the second phase, the MCS to be used for each sub-channel j is chosen based on the post-processing SINR given by

$$\gamma_{\text{post},j}^{\text{coopTxDiv2}} = \gamma_{SD,j} + \gamma_{RD,j}. \tag{4.2}$$

When the $S \rightarrow R$ link is reliable, this independent MCS selection in each phase help reduce the multiplexing loss caused by the two phased transmission nature of wireless relaying. This is due to the fact that, when the $S \rightarrow R$ link is reliable, the duration of the first phase can be shorter than that of the second phase as a higher rate MCS can be used for the transmissions in the first phase. With such link adaptation at a sub-channel j , the end-to-end throughput per channel use is given by

$$\rho_j^{\text{coopTxDiv2}} = \frac{\rho(\gamma_{SR,j})\rho(\gamma_{\text{post},j}^{\text{coopTxDiv2}})}{R(\gamma_{SR,j}) + R(\gamma_{\text{post},j}^{\text{coopTxDiv2}})}. \tag{4.3}$$

When the relays are deployed at strategic locations in the cell, the end-to-end throughput with cooperative transmit diversity-2 can outperform that of cooperative diversity-1. This is due to the fact that, cooperative transmit diversity-2 can provide rate adaptive relaying, whereas cooperative transmit diversity-1 provides fixed rate relaying. On the other hand, cooperative transmit diversity-1 can outperform cooperative transmit diversity-2 at locations far away from both the BS and the RSs. This is due to the fact that, at those regions, the SINR gain becomes more important than compensating for the multiplexing loss as the SINRs in the links get lower.

4.4.1.3 Cooperative Receive Diversity

Even if this scheme can achieve the same post-processing SINR as that of cooperative transmit diversity-2 (as derived in Chapter 3), it suffers from a potentially higher multiplexing loss due to the need for fixed rate relaying. Hence, cooperative receive diversity cannot outperform cooperative transmit diversity-2 which can provide rate adaptive relaying.

4.4.1.4 Cooperative Selection Diversity

When the CSI of the instantaneous SINR conditions is available at the BS, cooperative selection diversity can be achieved when the BS dynamically chooses to relay or not. In cooperative selection diversity, when relaying is decided, rate adaptive conventional relaying is used. Cooperative selection diversity is the least complex cooperative diversity scheme as it does not require any coherent combining at the MS. With conventional relaying, the post-processing SINR at the MS is given by $\gamma_{RD,j}$. For the first phase of conventional relaying, the MCS to be used is selected based on $\gamma_{SR,j}$ and for the second phase based on $\gamma_{RD,j}$. Hence, the end-to-end throughput with conventional relaying is given by

$$\rho_j^{\text{conv}} = \frac{\rho(\gamma_{SR,j})\rho(\gamma_{RD,j})}{R(\gamma_{SR,j}) + R(\gamma_{RD,j})}. \tag{4.4}$$

For a given user, the end-to-end throughput with cooperative selection diversity is then given by

$$\rho_j^{\text{coopSDiv}} = \max\{\rho_j^{\text{conv}}, \rho(\gamma_{SD,j})\}. \tag{4.5}$$

When relaying improves the performance as compared to that of w/o relay, the cooperative transmit diversity-2 can outperform the cooperative selection diversity as it can provide higher post-processing SINR. On the other hand, when the channel conditions in the $R \rightarrow D$ link is very good (i.e., $\gamma_{RD,j} \gg \gamma_{SD,j}$), the cooperative selection diversity can perform as well as cooperative transmit diversity-2. Due to the fact that cooperative transmit diversity-2 and conventional relaying enable rate adaptive relaying, they can outperform the other schemes which necessitate fixed rate relaying.

4.4.1.5 Adaptive Cooperative Diversity Scheme

Adaptive cooperative diversity scheme chooses (i) w/o relay transmissions if any of the afore-mentioned cooperative diversity schemes cannot outperform the w/o relay scheme, (ii) conventional relaying if it can outperform the w/o relay scheme and can perform as well as more complex cooperative diversity schemes which require combining at the MS, or (iii) the best scheme among cooperative transmit diversity-1 and cooperative transmit diversity-2 schemes if they can improve the end-to-end throughput as compared to both w/o relay and conventional relaying. Such an adaptive scheme introduces cooperative diversity which requires signal combining at the MS only when necessary and uses the scheme with the highest end-to-end throughput. Hence, it can reduce complexity while maximizing the end-to-end throughput. For a given user, the end-to-end throughput with adaptive cooperative diversity is then given by

$$\rho_j^{\text{AdaptiveCoopDiv}} = \max\left\{\rho_j^{\text{direct}}, \rho_j^{\text{conv}}, \rho_j^{\text{coopTxDiv2}}, \rho_j^{\text{coopTxDiv1}}\right\}, \tag{4.6}$$

where $\rho_j^{\text{direct}} = \rho(\gamma_{SD,j})$. If the two schemes have the same performance, the one with less complexity is selected. As per the input output equations derived in Chapter 3, the schemes with increasing complexity can be listed as follows: direct transmission, conventional relaying, cooperative transmit diversity-2 and cooperative transmit diversity-1 [2].

4.4.2 Scheduling of the Users in a Two-hop Cellular Network

Scheduling of the users in multi-hop cellular networks needs modifications on conventional scheduling algorithms designed for the w/o relay networks. This is due to the fact that, the end-to-end performance should be considered with efficient MCS in each hop rather than only SINR[1] or individual link throughput performance. For all the aforementioned cooperative diversity schemes in the previous section, the schedulers developed in this chapter use relaying only when the end-to-end throughput with the relay is greater than that of w/o relay.

In this chapter, the scheduling of the users on the radio resources for each DL frame $k \in \mathbb{N}$ is developed as the following. The scheduler at the BS uses a look-up table developed for point-to-point AWGN links with given SINR conditions. For each instantaneous SINR γ with resolution of 0.1 dB, the look-up table stores the throughput $\rho(\gamma)$ and the MCS which provides the highest throughput. This look-up table was introduced in Chapter 3. For each sub-channel j and for each user (i.e., MS) u, the scheduler calculates the post-processing SINR with the relay, i.e., $\gamma_{u,j}^{\text{post}}$. Let $\gamma_{SD,u,j}$ denote the instant-aneous SINR the user u experiences on sub-channel j in the $S \to D$ link. The scheduler plugs in $\gamma_{SD,u,j}$, $\gamma_{SR,j}$ and $\gamma_{u,j}^{\text{post}}$ to the look-up table and reads the corresponding throughput and nominal rate for each of them. Then, for each cooperative diversity scheme under consideration, it calculates the end-to-end throughput with the relay, i.e., $\rho_{u,j}^{\text{with}-\text{relay}}$, by the end-to-end throughput equations presented in Section 4.4.1. Let $\rho_{u,j}^{\text{direct}} = \rho(\gamma_{SD,u,j})$ denote the throughput that user u can obtain on sub-channel j without relay, i.e., via $S \to D$ link only. For each user and for each sub-channel, the scheduler first decides on to relay or not by

$$\rho_{u,j} = \max\{\rho_{u,j}^{\text{direct}}, \rho_{u,j}^{\text{with}-\text{relay}}\}. \tag{4.7}$$

---

[1] For instance, max SINR scheduler considers only SINR conditions and schedules the users with the best SINR conditions.

If $\rho_{u,j} = \rho_{u,j}^{\text{direct}}$, w/o relay transmission is chosen over sub-channel j for user u. In this case, sub-channel j carries one access burst spanning over the first and second phases. If $\rho_{u,j} = \rho_{u,j}^{\text{with-relay}}$, one of the relayed transmission is chosen. In this case, sub-channel j carries one relay burst in the first phase, and one access burst in the second phase. After the best scheme (with or w/o relay) is selected as above, the users are then scheduled according to various scheduling algorithms developed for two hop cellular networks. Such scheduling algorithms are presented in the following.

4.4.2.1 Maximum End-to-End Throughput Scheduler

At a given frame and sub-channel of a w/o relay system, the maximum throughput scheduler allocates the sub-channel to the user who achieves the maximum throughput. This is equivalent to scheduling the user who sees a peak SINR in the sub-channel. Maximum throughput scheduler provides multi-user diversity and maximizes the system throughput. However, it cannot provide fairness to the users as the users with good channel conditions will be scheduled more than the users with relatively worse channel conditions. In this section, the maximum throughput scheduler used in single hop cellular networks is modified to be efficiently used for two-hop cellular networks. This modified maximum throughput scheduler is referred to as maximum end-to-end throughput scheduler and it allocates the sub-channel j to the user $\hat{u}$ according to:

$$\hat{u} = \text{argmax}\,\{\rho_{u,j}\}. \tag{4.8}$$

Such scheduler will provide similar advantages as maximum throughput scheduler.

4.4.2.2 Proportional Fair Scheduling for a Two Hop Cellular Network

With Proportional Fair Scheduler (PFS) in wireless networks, the system can benefit from multi-user diversity while offering fairness to the users [59]. To obtain these benefits, the system sacrifices from the system throughput that can be achieved with the maximum end-to-end throughput scheduler. In this section, the existing PFS used in single-hop cellular networks is modified to be efficiently used for two-hop cellular networks. For each sub-channel j, the modified PFS calculates the PFS metric for each user according to [59][2]

$$\text{PFS}_{u,j} = \frac{\rho_{u,j}}{\overline{\rho}_u[k-1]}. \tag{4.9}$$

----

[2] Park et al. analyze the PFS for single-hop wireless networks in [59].

Then, it allocates the sub-channel j to the user $\hat{u}$ according to [59]

$$\hat{u} = \text{argmax}\, \{PFS_{u,j}\}. \tag{4.10}$$

The term $\overline{\rho}_u[k-1]$ represents the past average throughput of user u at DL frame $k-1$. Once the users are scheduled, the past average throughput for each user is updated by using a low pass filter with a time constant of T slots. This update is done according to

$$\overline{\rho}_u[k] = \frac{(T-1)\overline{\rho}_u[k-1] + \sum_{j=1}^{J}(c_{u,j}\rho_{u,j})}{T}. \tag{4.11}$$

The term $c_{u,j}$ is equal to one if user u is scheduled on sub-channel j, otherwise it is equal to zero. The designed scheduler provides both cooperative diversity and multi-user diversity. For a given user, it guarantees that the end-to-end throughput will always be greater than or equal to that of w/o relay. The time constant T adjusts the level of fairness of the scheduler. When $T = 1$, the users cannot get a fair channel access. Hence, T should be long enough to provide fairness to the users. In this chapter, the evaluations are provided with various T which reflects the proportional fair and proportional unfair scheduling cases.

4.5 Deployment of the Relays in the Cell

Firs of all, the relays should be deployed at strategic positions in the cell where LOS condition in the $S \rightarrow R$ links is possible. In this chapter, all the relays are positioned symmetrically at a distance of 10.4 km to the BS. With the current simulation set-up, the reason for this placement is threefold:

- First of all, this is the longest distance where it is possible to detect 64-QAM symbols with negligible error rate at each RS.
- Second, this is a far enough distance such that within the coverage of a given relay, the transmissions with the relay improves the performance as compared to w/o relay. For a single user, Figure 4.1 presents the average throughput gain per position in the cell. Figure 4.1 presents the throughput gain in terms of

$$\frac{(\overline{\rho}^{\text{coopTxDiv2}} - \overline{\rho}^{\text{direct}})}{\overline{\rho}^{\text{direct}}} \times 100.$$

The figure shows that the relays are positioned at strategic locations in the cell where the coverage of each relay is utilized efficiently. The gains

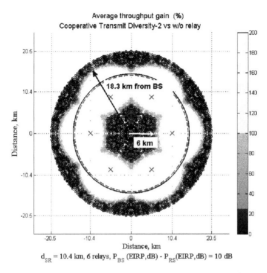

$d_{SR} = 10.4$ km, 6 relays, P_{BS} (EIRP,dB) - P_{RS}(EIRP,dB) = 10 dB

Figure 4.1 Average throughput gain per position in the cell. The gain is presented as a percentage and shows the advantage of dynamically using *the best scheme among cooperative transmit diversity-2 and w/o relay* instead of w/o relay only. A full colour version is available on http://www.researchwebshelf.com/DocuDetails.php?SrlNo=46.

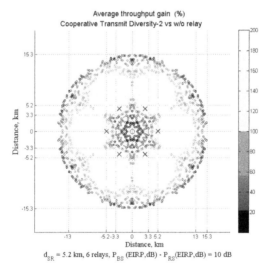

$d_{SR} = 5.2$ km, 6 relays, P_{BS} (EIRP,dB) - P_{RS}(EIRP,dB) = 10 dB

Figure 4.2 Average throughput gain per position in the cell. The gain is presented as a percentage and shows the advantage of dynamically using *the best scheme among cooperative transmit diversity-2 and w/o relay* instead of w/o relay only. The relays are positioned at a distance of 5.2 km from the BS. A full colour version is available on http://www.researchwebshelf.com/DocuDetails.php?SrlNo=46.

greater than 200% are rounded down to 200%. The gain at close distances to the BS, i.e., up to 6 km, is zero as in those regions transmissions w/o relay provide the highest end-to-end throughput. If the relays were placed closer to the BS for example, then some part of the relays' coverage will be wasted as in those parts the transmission with the relay would not improve the end-to-end performance. To justify this statement, the investigation presented in Figure 4.2 is conducted. The result in this figure is obtained for six relays which are positioned symmetrically at a distance of 5.2 km from the BS. A single user is simulated to see the relative performance versus SNR conditions at various positions in the cell. Figure 4.2 presents the average throughput gain per position in the cell in terms of

$$\frac{(\overline{\rho}^{\text{coopTxDiv2}} - \overline{\rho}^{\text{direct}})}{\overline{\rho}^{\text{direct}}} \times 100.$$

The gains greater than 200% are rounded down to 200%. The figure shows that the relays are not positioned at strategic locations in the cell due to the fact that the potential coverage area of a given relay is not used efficiently. When the relays are positioned at a distance of 5.2 km from the BS, they are within the area which corresponds to the coverage area the w/o relay system.[3] Within the area which corresponds to the coverage area of the w/o relay system, even if a user sees a good channel condition to the RS, it does not need relaying most of the time. Therefore, in order to efficiently use the potential coverage area of the relays, they should be deployed outside the area which corresponds to the coverage area of the w/o relay system.

- Third, there is a region where the SNRs in the $S \to D$ and $R \to D$ links are comparable to each other where cooperative diversity is beneficial. Over such region, the users can get service either from the BS or RS, whichever is beneficial. Such an SNR region can be seen in Figure 4.3. In Figure 4.3, the ratio of $\text{SNR}_{SD}/\text{SNR}_{RD}$ is plotted in dB for each position in the cell. In the figure, the results at a range of up to 20.5 km are plotted for investigation.[4] The relays could have been placed farther from the BS while allowing decoding errors at the relays. However, such decoding errors in the $S \to R$ links will severely limit the end-to-end throughput performance of wireless relaying, since it is

---

[3] The w/o relay system with the PFS (with $T = 100$) has a coverage area of 8.4 km as will be shown later in this chapter.

[4] Note that the cell radius is not necessarily 20.5 km.

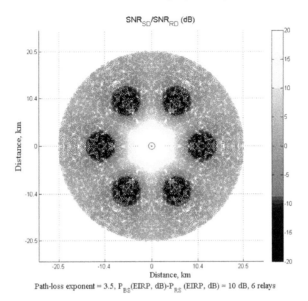

Figure 4.3 The long-term average SNR difference in dB per position in the cell. A full colour version is available on http://www.researchwebshelf.com/DocuDetails.php?SrlNo=46.

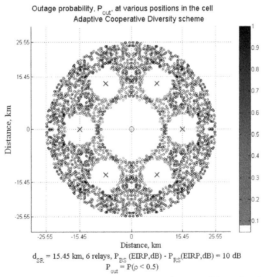

Figure 4.4 The outage probability, i.e. $1 - p$ at various positions in the cell. Within the white regions, users are within the coverage area. A full colour version is available on http://www.researchwebshelf.com/DocuDetails.php?SrlNo=46.

already limited by the multiplexing loss. Furthermore, if the relays were placed at farther distances to the BS, then the SNR difference between the signals received from the relay and the BS increases. This in turn diminishes the benefits of cooperative diversity. This might cause some regions to be out of coverage (i.e., in outage). Such effect is visualized in Figure 4.4 for a total of 60 users uniformly distributed in the cell. Figure 4.4 presents the outage probability at various positions in the cell. To obtain the results, the PFS developed in this chapter is used with the time constant T set to 100 to provide fairness to the users. The outage probability is calculated with the most robust scheme, i.e., the adaptive cooperative diversity scheme. The result in this figure is plotted for 6 relays which are positioned at 15.45 km away from the BS. Within the white regions, users are within the coverage area. Outside of these white regions, $P_{out} > 0.05$ and hence the users are in outage. This shows that placing the relays at further away distances to the BS results in areas which are in outage.

4.6 The Relative Performance Evaluation of the Cooperative Diversity Schemes with Maximum End-to-End Throughput Scheduler

In this section, the relative performance of various cooperative diversity schemes is presented with the maximum end-to-end scheduler. To obtain the results, a total of 60 users are positioned uniformly within a region with a minimum BS, MS distance of 6 km, and a maximum BS, MS distance of 14.85 km. In Figure 4.5, the average number of sub-channels allocated to a given user is plotted at various positions in the cell. In Figure 4.6, the average throughput per channel use is plotted at various positions in the cell. In Figures 4.5 and 4.6, a w/o relay system has been analyzed. As the results show, the users that are closer to the BS can obtain more sub-channels than the users that are farther to the BS. The users within a distance of 6.4 km to the BS are in coverage area with $p \geq 0.95$. In Figure 4.7, the average end-to-end throughput is plotted at various positions in the cell. The cooperative-transmit diversity-2 scheme is analyzed in this figure. The relays are positioned at a distance of 10.4 km to the BS. The existence of unoccupied regions (i.e., the regions in white) shows that the users only at close distances to the BS and a given RS can obtain a sub-channel. This shows the unfairness of the maximum end-to-end throughput scheduler. However, it

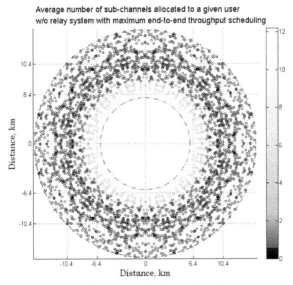

Figure 4.5 The average number of sub-channels allocated to the users at various positions in the cell. A full colour version is available on http://www.researchwebshelf.com/DocuDetails.php?SrlNo=46.

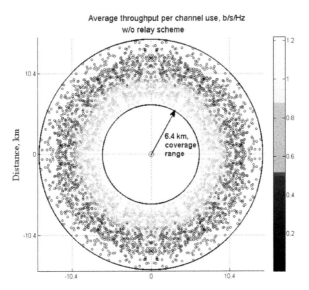

Figure 4.6 The average throughput per channel use in a w/o relay system with maximum end-to-end throughput scheduler. A full colour version is available on http://www.researchwebshelf.com/DocuDetails.php?SrlNo=46.

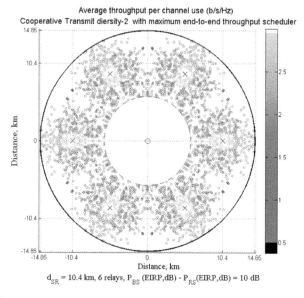

Figure 4.7 The average throughput per channel use at various positions in the cell with cooperative transmit diversity-2 scheme. A full colour version is available on http://www.researchwebshelf.com/DocuDetails.php?SrlNo=46.

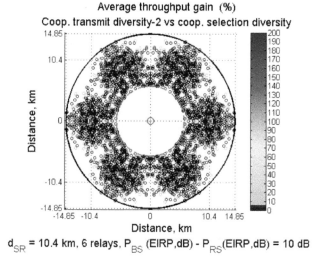

Figure 4.8 Average throughput gain per position in the cell. The gain is presented as a percentage and shows the advantage of dynamically using *the best scheme among cooperative transmit diversity 2 and w/o relay* as compared to cooperative selection diversity. A full colour version is available on http://www.researchwebshelf.com/DocuDetails.php?SrlNo=46.

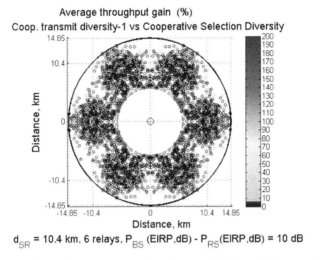

Figure 4.9 Average throughput gain per position in the cell. The gain is presented as a percentage and shows the advantage of dynamically using *the best scheme among cooperative transmit diversity 1 and w/o relay* as compared to cooperative selection diversity. A full colour version is available on http://www.researchwebshelf.com/DocuDetails.php?SrlNo=46.

maximizes the average end-to-end throughput in the cell by allocating the sub-channels to the users who see a peak end-to-end throughput. In Figure 4.8, $(\overline{\rho}^{\text{coopTxDiv2}} - \overline{\rho}^{\text{coopSDiv}})/\overline{\rho}^{\text{coopSDiv}} \times 100$ is plotted per position in the cell. In Figure 4.9, $(\overline{\rho}^{\text{coopTxDiv1}} - \overline{\rho}^{\text{coopSDiv}})/\overline{\rho}^{\text{coopSDiv}} \times 100$ is plotted per position in the cell. As the results in Figures 4.8 and 4.9 show, only the users that are relatively closer to a given RS and BS can obtain a sub-channel. Hence, cooperative transmit diversity-1 and cooperative transmit diversity-2 schemes cannot provide significant throughput gain as compared to the cooperative selection diversity. Consequently, with the maximum end-to-end throughput scheduler, the use of adaptive relaying scheme will not bring throughput enhancement as compared to using cooperative selection diversity only.

4.7 The Relative Performance Evaluation of the Cooperative Diversity Schemes with PFS

In this section, the relative performance evaluation of various cooperative diversity schemes is presented with the PFS developed in this chapter. Two cases have been analyzed. The case where the users can obtain a fair channel

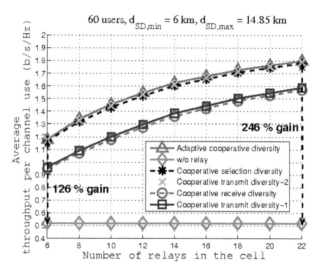

Figure 4.10 Average throughput per channel use versus the number of relay stations in the cell. For each cooperative diversity scheme, the users are scheduled with the best scheme among w/o relay and with relay. The minimum and maximum distance of the users to the BS are $d_{SD,min} = 6$ km and $d_{SD,max} = 14.85$ km, respectively. A full colour version is available on http://www.researchwebshelf.com/DocuDetails.php?SrlNo=46.

access and the case where the users can obtain a less fair channel access have been analyzed. To obtain these cases, the fairness is controlled by adjusting the time constant T.

4.7.1 The Evaluation with High Fairness

In this section, the relative performance evaluation of various cooperative diversity schemes is presented for $T = 100$. Such time constant provides equal channel access chance to the users regardless of their position in the cell. In the figures, within the black circle with radius 14.85 km, $p \geq 0.7$ with adaptive cooperative diversity. Within the dashed circle with radius 14.5 km, $p \geq 0.95$ with cooperative-selection diversity. For $p \geq 0.7$, the adaptive cooperative diversity provides the highest coverage area. For $p \geq 0.95$, the cooperative selection diversity is providing the lowest coverage area among all the other cooperative diversity schemes. For $p \geq 0.95$, the w/o relay system provides a coverage area with a radius of $r = 8.4$ km.

In Figure 4.10, the overall average throughput per channel use is plotted versus the total number of relays in the cell. A total of 60 users have been

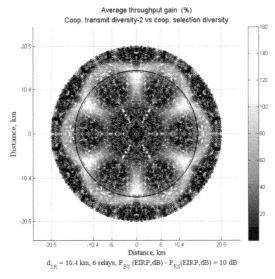

Figure 4.11 Average throughput gain per position in the cell. The gain is presented as a percentage and shows the advantage of dynamically using *the best scheme among cooperative transmit diversity-2 and w/o relay* as compared to cooperative selection diversity. A full colour version is available on http://www.researchwebshelf.com/DocuDetails.php?SrlNo=46.

simulated to obtain the results in this figure. The average throughput is calculated within a range greater than 6 km and smaller than 14.85 km to the BS. This is the area where relaying improves the performance as compared to the w/o relay system (see also Figure 4.1) and where the users are within the coverage area. As seen in the figure, the cooperative selection diversity can perform as well as more complex cooperative diversity schemes such as adaptive cooperative diversity and cooperative transmit diversity-2. When there are a total of six relays in the cell, the cooperative selection diversity brings a throughput enhancement of 126% as compared to the w/o relay transmissions. This gain increases with the increasing number of relays deployed in the cell. The throughput gain that can be achieved with cooperative selection diversity reaches 246% at a number of relays equal to 22. In the system level average throughput sense, cooperative transmit diversity-2 and cooperative selection diversity outperform the cooperative transmit diversity-1 even if it can provide highest post-processing SINR. The reasons for the conclusions drawn from Figure 4.10 can further be explained with the investigations presented in Figures 4.11 and 4.12.

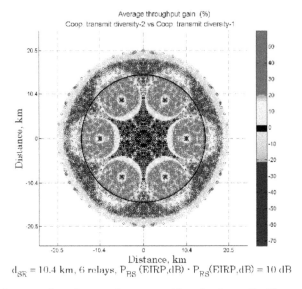

Figure 4.12 Average throughput gain per position in the cell. The gain is presented as a percentage and shows the advantage of dynamically using *the best scheme
among cooperative transmit diversity-2 and w/o relay* instead of *the best scheme among
cooperative transmit diversity-1 and w/o relay*. A full colour version is available on
http://www.researchwebshelf.com/DocuDetails.php?SrlNo=46.

For a single user, Figure 4.11 presents the average throughput gain at
various positions in the cell. In Figure 4.11, the throughput gain per position is
calculated by $(\overline{\rho}^{\text{coopTxDiv2}} - \overline{\rho}^{\text{coopSDiv}})/\overline{\rho}^{\text{coopSDiv}} \times 100$. As seen in this figure,
cooperative transmit diversity-2 can bring a throughput gain of around 25%
in the majority of the region where it is beneficial over conventional relaying.
Such a gain can be seen over the region where the received SNR from the BS
and the closest RS are comparable to each other. Over that region, the benefits of increased post-processing SNR as compared to conventional relaying
becomes important. However, such a region is only a small fraction of the
overall coverage area and hence, the overall gain in the average throughput
becomes insignificant. Furthermore, as the number of relays is increased in
order to increase the throughput, such a region gets smaller.

For a single user, Figure 4.12 presents the throughput gain of
cooperative transmit diversity-2 as compared to cooperative transmit diversity-1 at various positions in the cell. In Figure 4.12,
$(\overline{\rho}^{\text{coopTxDiv2}} - \overline{\rho}^{\text{coopTxDiv1}})/\overline{\rho}^{\text{coopTxDiv1}} \times 100$ is plotted per position in the cell.
As the figure shows, cooperative transmit diversity-1 can outperform the coo-

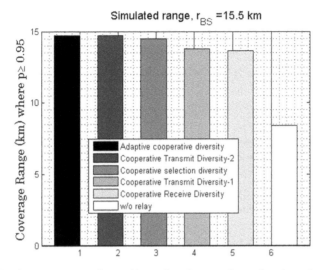

Figure 4.13 The coverage radius with each scheme where the throughput for each user is greater than 0.5 b/s/Hz with probability greater than 0.95. For each cooperative diversity scheme, the relay is used only if it can improve the end-to-end throughput as compared to w/o relay. A full colour version is available on http://www.researchwebshelf.com/DocuDetails.php?SrlNo=46.

perative transmit diversity-2 when the user is far from both the BS and the closest RS. Due to the high path loss in such a region, the SNR gain becomes more important than the multiplexing loss. On the other hand, such a region is small and outside of the coverage area. Hence, in terms of average throughput and coverage, cooperative transmit diversity-1 cannot provide significant gain as compared to cooperative transmit diversity-2. Therefore, adaptive cooperative diversity performs similar to cooperative transmit diversity-2 and cooperative selection diversity.

Figure 4.13 presents the coverage radius r (originated from the BS) achieved with each one of the cooperative diversity schemes analyzed in this chapter. The conclusions drawn from this figure are similar to that drawn from Figure 4.10. The w/o relay system provides a coverage range of 8.4 km. The cooperative transmit diversity-2 and adaptive cooperative diversity schemes provide the same coverage range which is 14.7 km. The cooperative selection diversity provides negligibly (1.4%) less coverage range as compared to that provided by cooperative transmit diversity-2. As seen in Figure 4.13, the cooperative selection diversity can provide better coverage as compared to both cooperative receive diversity and cooperative transmit diversity-1.

Cooperative transmit diversity-1 and cooperative receive diversity provide similar coverage range of 13.8 and 13.65 km, respectively. This is due to the following. As compared to both cooperative receive diversity and cooperative transmit diversity-2, cooperative transmit diversity-1 can increase the post-processing SINR as much as $\gamma_{SD,j}$ for each sub-channel. Such SINR gain can provide significant throughput increase at closer distances to the BS. However, over closer distances to the BS, the users are scheduled without relay most of the time and they are already within the coverage area. Hence, such SINR gain becomes important only at far distances to the BS where relaying is beneficial. Over such region, SINR in the $S \rightarrow D$ link is already low. Hence, for two-hop cellular networks with fixed relays, the cooperative transmit diversity-1 cannot bring significant advantage in the average sense as compared to cooperative receive diversity and cooperative transmit diversity-2. With the cooperative transmit diversity-2, the coverage area of an existing IEEE 802.16e system can be tripled via the extension to the IEEE 802.16j standard with the scheduler developed herein. Hence, this will reduce the deployment costs of two out of three BSs. However, to achieve this saving, several relay stations, e.g. six, need to be deployed. Hence, to achieve savings from the overall deployment costs, the cost of an RS needs to be less than one third of the cost of a BS if a total of six relays are deployed in a cell. If the relays are designed properly with basic transceiver structures with only PHY and MAC functionality, the infrastructure cost of an RS can be much less than that of a BS.

4.7.2 The Evaluation with Lower Fairness

In this section, the relative performance evaluation of various cooperative diversity schemes is given for $T = 5$. Such time constant does not provide equal channel access chance to the users. It will favor users closer to the BS or a given RS more than the users that are farther away to the BS or a given RS. This results in increased cell throughput and decreased coverage range as compared to a fairer PFS. When T decreases, the performance of PFS approaches to that of maximum end-to-end throughput scheduler. Figure 4.14 presents the coverage range (originating from the location of the BS) achieved with each one of the cooperative diversity schemes analyzed in this chapter. The conclusions drawn from this figure are similar to those drawn from Figure 4.13. The w/o relay system can achieve a coverage range of 7 km which is less than that can be achieved with the PFS with $T = 100$. The cooperative transmit diversity-2 and adaptive cooperative diversity schemes provide the

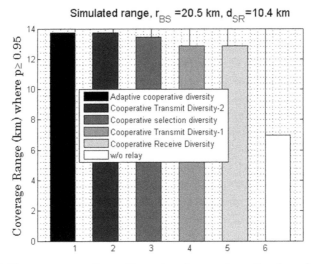

Figure 4.14 The coverage radius with each scheme where the throughput for each user is greater than 0.5 b/s/Hz with probability greater than 0.95. For each cooperative diversity scheme, the relay is used only if it can improve the end-to-end throughput as compared to w/o relay. A full colour version is available on http://www.researchwebshelf.com/DocuDetails.php?SrlNo=46.

same coverage range which is 13.7 km. The cooperative selection diversity provides negligibly (2%) less coverage range as compared to that provided by cooperative transmit diversity-2. When T is decreased, Figure 4.14 shows that the coverage range for each scheme reduces. The scheduled users tend to be closer to a given relay. Hence, cooperative transmit diversity-2 and adaptive cooperative diversity schemes will not provide significant performance gain as compared to cooperative-selection diversity.

4.8 Comparative Analysis of the End-to-End Maximum Achievable Rate of Various Relaying Schemes

After developing efficient link adaptive transmission schemes for implementation in two-hop wireless relay networks, this section provides information theoretic relative performance evaluation of various relaying schemes. The conclusions provide guidelines for determining the most efficient relaying and forwarding schemes from an information theoretic perspective.

4.8.1 Introduction

The channel capacity in b/s/Hz is a measure of the maximum achievable data rate per unit bandwidth that can be communicated to the destination terminal with negligible error rate [10, 34]. The channel throughput in b/s/Hz can be defined as the percentage of the data rate per unit bandwidth that is communicated to the destination terminal without error. The capacity analysis yields insights into the maximum achievable throughput by the system.

The study in [5] compares the ergodic capacity of Cooperative-MIMO, Cooperative-SIMO and Cooperative-MISO schemes and concentrates on frequency flat environments. For each relaying scheme, it assumes two phases with equal duration, i.e., fixed rate relaying, which is not optimal for cooperative-MISO and conventional-relaying schemes as shown in Chapters 3 and 4. Hence, the conclusions in [5] underestimate the performance of cooperative-MISO scheme. The investigation in [5] does not contain any comparisons to w/o relay transmission and conventional relaying. Furthermore, the comparisons are provided only when the SNR condition in the direct link is high enough such that relaying may not improve the system performance.[5] Consequently, the comparisons presented in [5] did not show the relative performance when relaying is beneficial. The study in [5] does not provide performance comparison of AF and DF-based relaying. Only fixed relaying, where one type of relaying and forwarding is used at all times, is analyzed in [5]. The study in [60] provides a capacity analysis of OFDM and OFDMA-based wireless relay networks. It provides only a capacity analysis of the Cooperative-SIMO scheme and does not contain any comparison with other relaying mechanisms.

The analysis presented in this chapter provides a unified framework which compares the end-to-end achievable rate of various relaying schemes over the region where relaying improves the performance. The achievable rates with AF and DF-based relaying are analyzed and compared to each other as well. The comparisons are provided for various channel conditions in a relay network. Furthermore, a fully adaptive relaying scheme which dynamically uses the best relaying scheme and decides whether to relay or not is analyzed in terms of information theoretic performance. The performance of such adaptive relaying scheme is compared to its simpler versions, where one type of relaying scheme is used if relaying is beneficial. This information theoretic analysis is necessary to validate the conclusions drawn from an implementa-

---

[5] This is verified with the results presented in the following sections.

tion point of view which are presented in Chapter 3 and the previous sections of this chapter.

4.8.2 The End-to-End Maximum Achievable Rate with Various Cooperative Multi-Antenna Channels

In this section, the end-to-end maximum achievable rate of various relaying schemes is derived with the system considerations outlined in Section 2.3. The relaying schemes outlined in Table 2.1 and w/o relay scheme are considered. The derivations are provided for transmissions taking place at a given subcarrier i scheduled to a given user. The derivations are provided for both AF and DF-based forwarding. At a given subcarrier i, the maximum achievable rate of w/o relay scheme is given by [61]

$$C_{\text{direct}} = \log_2(1 + \gamma_{SD,i}) \text{ b/s/Hz.} \tag{4.12}$$

4.8.2.1 AF-Based Relaying

In this section, the cooperative multi-antenna channels are created via AF-based relaying. In the following, ζ_i is defined as

$$\zeta_i = 1 + \frac{\gamma_{RD,i}}{(1 + \gamma_{SR,i})}. \tag{4.13}$$

1. *Cooperative-MIMO Channel*: Let $x_{1,i}$ and $x_{2,i}$ represent the transmitted constellation points by the BS at the first and second phase, respectively. Let $y_{R,i}[m]$ and $y_{D,i}[m]$ represent the received signals at the relay and destination terminals at the first phase. $y_{R,i}[m]$ and $y_{D,i}[m]$ are given by

$$y_{R,i}[m] = h_{SR,i}\sqrt{E_S}x_{1,i} + n_{R,i}[m], \tag{4.14}$$

$$y_{D,i}[m] = h_{SD,i}\sqrt{E_S}x_{1,i} + n_{D,i}[m]. \tag{4.15}$$

With a cooperative-MIMO scheme, let $y_{D,i}[m + 1]$ represent the received signal at destination terminal at the second phase. $y_{D,i}[m + 1]$ is given by

$$
\begin{aligned}
y_{D,i}[m + 1] &= h_{SD,i}\sqrt{E_S}x_{2,i} + h_{RD,i}\frac{\sqrt{E_R}}{\beta_i}y_{R,i}[m] + n_{D,i}[m + 1] \\
&= h_{SD,i}\sqrt{E_S}x_{2,i} + \frac{\sqrt{E_R E_S}}{\beta_i}h_{SR,i}h_{RD,i}x_{1,i} \\
&\quad + h_{RD,i}\frac{\sqrt{E_R}}{\beta_i}n_{R,i}[m] + n_{D,i}[m + 1].
\end{aligned}
\tag{4.16}
$$

The term β_i is defined in Equation (2.4). Let n_2 represent the noise component in $y_{D,i}[m+1]$. n_2 is given by

$$n_2 \triangleq n_{D,i}[m+1] + h_{RD,i}\frac{\sqrt{E_R}}{\beta_i}n_{R,i}[m]. \tag{4.17}$$

The power of the noise signal at the second phase can be calculated as

$$\mathcal{E}\{|n_2|^2\} = N_o^D\left(1 + \frac{\gamma_{RD,i}}{(1+\gamma_{SR,i})}\right) = N_o^D\zeta_i. \tag{4.18}$$

n_2 is a Zero Mean Circularly Symmetric Complex Gaussian (ZMCSCG) random variable, i.e., $n_2 \sim \mathcal{CN}(0, \mathcal{E}\{|n_2|^2\})$.

The effective input-output relation at subcarrier i over the two phases can be written as

$$\mathbf{y} = \begin{bmatrix} y_{D,i}[m] \\ y_{D,i}[m+1] \end{bmatrix} = \begin{bmatrix} h_{1,1} & h_{1,2} \\ h_{2,1} & h_{2,2} \end{bmatrix}\begin{bmatrix} x_{1,i} \\ x_{2,i} \end{bmatrix} + \begin{bmatrix} n_1 \\ n_2 \end{bmatrix}$$
$$= \mathbf{H}^{\mathrm{AF}}\mathbf{x} + \mathbf{n}, \tag{4.19}$$

where the 2×2 cooperative-MIMO channel $\mathbf{H}^{\mathrm{AF}}$ is given by Equation (3.1). The term n_1 is defined as; $n_1 \triangleq n_{D,i}[m]$. The noise vector received during two phases is given by $\mathbf{n} = [n_1 \ n_2]^T$. The term $\mathbf{x}$ is given by $\mathbf{x} = [x_{1,i} \ x_{2,i}]^T$. The covariance matrix of noise vector $\mathbf{n}$ is derived as

$$\mathcal{E}\{\mathbf{n}\mathbf{n}^H\} = \mathbf{R_{nn}} = \begin{bmatrix} N_o^D & 0 \\ 0 & N_o^D(1 + \frac{\gamma_{RD,i}}{(1+\gamma_{SR,i})}) \end{bmatrix}. \tag{4.20}$$

The vector $\mathbf{n}$ is ZMCSCG [62]. Since $\mathbf{n}$ is ZMCSCG, the differential entropy of vector $\mathbf{n}$ is given by [62]

$$H(\mathbf{n}) = \frac{1}{2}\log_2(\det(\pi e\mathbf{R_{nn}})) \ \mathrm{b/s/Hz}. \tag{4.21}$$

Since $\mathbf{x}$ and $\mathbf{n}$ are independent and have zero mean, the covariance matrix of $\mathbf{y}$ is given by

$$\mathcal{E}\{\mathbf{y}\mathbf{y}^H\} = \mathbf{R_{yy}} = \mathbf{H}^{\mathrm{AF}}\mathbf{R_{xx}}\mathbf{H}^{\mathrm{AF}H} + \mathbf{R_{nn}}. \tag{4.22}$$

The mutual information between vectors $\mathbf{x}$ and $\mathbf{y}$ is given by [10]

$$I_{\mathbf{x};\mathbf{y}} = H(\mathbf{y}) - H(\mathbf{n}). \tag{4.23}$$

In this study it is assumed that transmitters do not know $\mathbf{H}^{AF}$ and destination terminal has the knowledge of $\mathbf{H}^{AF}$. The differential entropy $H(\mathbf{y})$ and hence $I_{\mathbf{x};\mathbf{y}}$ is maximized when $\mathbf{y}$ is ZMCSCG [62]. This in turn implies that $\mathbf{x}$ must be ZMCSCG with $\mathbf{R_{xx}} = \mathbf{I}_2$. In this case, the differential entropy of vector $\mathbf{y}$ is given by [62]

$$H(\mathbf{y}) = \frac{1}{2}\log_2(\det(\pi e \mathbf{R_{yy}})) \text{ b/s/Hz} \tag{4.24}$$

$$= \frac{1}{2}\log_2(\det(\pi e \mathbf{H}^{AF}\mathbf{H}^{AF^H} + \pi e \mathbf{R_{nn}})). \tag{4.25}$$

By using Equations (4.21), (4.23) and (4.25), the channel capacity of a cooperative-MIMO channel without the channel knowledge at the transmitters can be derived as

$$C_{Cooperative-MIMO}^{AF} = \frac{1}{2}\log_2\left(\frac{\det(\mathbf{H}^{AF}\mathbf{H}^{AF^H} + \mathbf{R_{nn}})}{\det(\mathbf{R_{nn}})}\right) \tag{4.26}$$

$$= \frac{1}{2}\log_2\left(1 + \frac{\gamma_{SD,i}^2}{\zeta_i} + \gamma_{SD,i} + \frac{\gamma_{RD,i}\gamma_{SR,i}}{\zeta_i(1+\gamma_{SR,i})} + \frac{\gamma_{SD,i}}{\zeta_i}\right).$$

2. *Cooperative-SIMO Channel*: The cooperative-SIMO channel created via AF-based relaying is given in Equation (2.6). The input-output relations that create a cooperative-SIMO channel at the destination terminal have been analyzed in Section 2.4.2 and are given by Equations (2.5), (2.6) and (2.7). The power of the noise signal at the second phase can be calculated as

$$\mathcal{E}\{|n_2|^2\} = N_o^D\left(1 + \frac{\gamma_{RD}}{1+\gamma_{SR}}\right). \tag{4.27}$$

The covariance matrix of the noise vector $\mathbf{n} = \mathbf{n}^{AF}$ can be calculated as

$$\mathcal{E}\{\mathbf{nn}^H\} = \mathbf{R_{nn}} = \begin{bmatrix} N_o^D & 0 \\ 0 & N_o^D(1 + \frac{\gamma_{RD}}{1+\gamma_{SR}}) \end{bmatrix}. \tag{4.28}$$

The vector $\mathbf{n}$ is ZMCSCG [62]. Since $\mathbf{n}$ is ZMCSCG, the differential entropy of vector $\mathbf{n}$ is given by Equation (4.21) [62].

Since the transmitted symbol $x_i[m]$ and the noise vector $\mathbf{n}$ are independent and have zero mean, the covariance matrix of $\mathbf{y}^{AF}$ is given by

$$\mathcal{E}\{\mathbf{y}^{AF}\mathbf{y}^{AF^H}\} = \mathbf{R_{yy}} = \mathbf{h}^{AF}R_{xx}\mathbf{h}^{AF^H} + \mathbf{R_{nn}}, \tag{4.29}$$

where R_{xx} is the covariance of the transmitted symbol $x_i[m]$. The mutual information between vectors $x = x_i[m]$ and $\mathbf{y}^{AF}$ is then given by [10]

$$I_{x;\mathbf{y}^{AF}} = H(\mathbf{y}^{AF}) - H(\mathbf{n}). \qquad (4.30)$$

It is assumed that transmitters do not know $\mathbf{h}^{AF}$ and destination terminal has the knowledge of $\mathbf{h}^{AF}$. The differential entropy $H(\mathbf{y}^{AF})$ and hence $I_{x;\mathbf{y}^{AF}}$ is maximized when $\mathbf{y}^{AF}$ is ZMCSCG [62]. This in turn implies that x must be ZMCSCG with $R_{xx} = 1$. In this case, the differential entropy of vector $\mathbf{y}^{AF}$ is given by [62]

$$H(\mathbf{y}^{AF}) = \frac{1}{2} \log_2(\det(\pi e \mathbf{R_{yy}})) \text{ b/s/Hz} \qquad (4.31)$$

$$= \frac{1}{2} \log_2(\det(\pi e \mathbf{h}^{AF}\mathbf{h}^{AF\mathbf{H}} + \pi e \mathbf{R_{nn}})). \qquad (4.32)$$

By using Equations (4.30), (4.32), (4.21) and (4.28), the channel capacity of a cooperative-SIMO channel without channel knowledge at the transmitters can be derived as

$$C^{AF}_{\text{Cooperative}-\text{SIMO}} = \frac{1}{2} \log_2 \left(\frac{\det(\mathbf{h}^{AF}\mathbf{h}^{AF\mathbf{H}} + \mathbf{R_{nn}})}{\det(\mathbf{R_{nn}})} \right) \qquad (4.33)$$

$$= \frac{1}{2} \log_2 \left(1 + \gamma_{SD,i} + \frac{\gamma_{RD,i}\gamma_{SR,i}}{\zeta_i(1 + \gamma_{SR,i})} \right). \qquad (4.34)$$

3. *Cooperative-MISO Channel*: The cooperative-MISO channel is a subset of cooperative-MIMO channel and it is given by Equation (3.13). The end-to-end maximum achievable rate with this Cooperative-MISO channel can be derived as

$$C^{AF}_{\text{Cooperative}-\text{MISO}} = \frac{1}{2} \log_2 \left(1 + \frac{\gamma_{RD,i}\gamma_{SR,i}}{(1 + \gamma_{SR,i})\zeta_i} + \frac{\gamma_{SD,i}}{\zeta_i} \right). \qquad (4.35)$$

4. *Conventional Relaying*: The end-to-end maximum achievable rate with conventional relaying can be derived as

$$C^{AF}_{\text{conv.}-\text{relaying}} = \frac{1}{2} \log_2 \left(1 + \frac{\gamma_{RD,i}\gamma_{SR,i}}{\zeta_i(1 + \gamma_{SR,i})} \right) \text{ b/s/Hz.} \qquad (4.36)$$

In the above derivations, the factor of $1/2$ in front of the end-to-end maximum achievable rate equations accounts for the fact that two phases with equal duration are needed in order to create the corresponding channel with AF.

4.8.2.2 DF-Based Relaying

In this section, the cooperative multi-antenna channels are created via DF-based relaying. In the following, $C_{SR} = \log_2(1+\gamma_{SR,i})$ represents the channel capacity of the subcarrier i of the $S \rightarrow R$ link. Let R_1 in b/s/Hz represent the transmission rate in the first phase and let R_2 in b/s/Hz represent the transmission rate in the second phase. In order to have error free detection at the RS and hence achieve capacity, R_1 should satisfy $R_1 \leq C_{SR}$. Assuming error free detection at the RS, the cooperative-MIMO channel achieved with DF-based relaying is given by Equation (3.2). The end-to-end maximum achievable sum rate via Cooperative-MIMO channel, $(R_1 + R_2)^{\text{max}}$, can be derived as [61]

$$(R_1 + R_2)^{\text{max}} = \frac{1}{2} \min\{(R_1^{\text{max}} + R_2^{\text{max}}), R_{\text{total}}\} \text{ b/s/Hz}, \qquad (4.37)$$

where

$$R_1^{\text{max}} = \min\{\log_2(1 + \gamma_{SD,i} + \gamma_{RD,i}), C_{SR}\},$$

$$R_2^{\text{max}} = \log_2(1 + \gamma_{SD,i}),$$

$$R_{\text{total}} = \log_2 \det\left(\mathbf{I}_2 + \frac{\mathbf{H}_{\text{MIMO}}^{\text{DF}}(\mathbf{H}_{\text{MIMO}}^{\text{DF}})^H}{N_o^D}\right)$$

$$= \log_2(1 + \gamma_{RD,i} + 2\gamma_{SD,i} + \gamma_{SD,i}^2).$$

The cooperative-SIMO channel is given by Equation (2.12). The end-to-end maximum achievable rate with the Cooperative-SIMO channel is given by

$$R_1^{max} = \frac{1}{2} \min\{C_{SR}, \log_2(1 + \gamma_{RD,i} + \gamma_{SD,i})\} \text{ b/s/Hz}. \qquad (4.38)$$

Note that for cooperative-MISO and conventional relaying schemes, the MS does not receive in the first phase. Hence, the rates in the first and second phase can be adjusted based on the channel seen at the receiving terminal in each phase, i.e at the RS in the first phase and at the MS in the second phase. Hence, the end-to-end achievable rates for these schemes can be derived as the following. The end-to-end maximum achievable rate via a cooperative-MISO scheme is given by

$$R^{max} = \frac{C_{SR} R_{\text{total}}}{(C_{SR} + R_{\text{total}})} \text{ b/s/Hz}, \qquad (4.39)$$

where $R_{\text{total}} = \log_2(1 + \gamma_{RD,i} + \gamma_{SD,i})$.

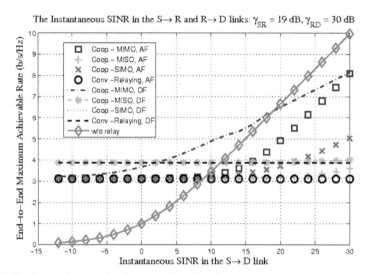

Figure 4.15 The maximum gain of cooperative-MIMO scheme as compared to other schemes.

The end-to-end maximum achievable rate via conventional relaying is given by

$$R^{\max} = \frac{C_{SR}C_{RD}}{(C_{SR} + C_{RD})} \text{ b/s/Hz}, \qquad (4.40)$$

where $C_{RD} = \log_2(1 + \gamma_{RD,i})$ b/s/Hz.

4.8.2.3 Performance Comparison

In this section, the information theoretic end-to-end maximum achievable rate of various relaying schemes are compared to each other. The SINR ranges considered in the analysis are $\gamma_{SR,i} \in [-12, 30]$ dB, $\gamma_{RD,i} \in [-12, 30]$ dB and $\gamma_{SD,i} \in [-12, 30]$ dB. The following conclusions are drawn over the SINR region where relaying improves the achievable rate as compared to that of a w/o relay system.

The analysis with the derived end-to-end maximum achievable rate equations show that AF-based relaying cannot outperform DF-based relaying for cooperative-MIMO, cooperative-MISO and conventional relaying schemes. With cooperative-SIMO scheme, AF-based relaying can outperform DF-based relaying as much as 0.16 b/s/Hz, which is insignificant for broadband applications. DF-based relaying on the other hand can provide significant performance improvement over AF-based relaying as much as 0.5 b/s/Hz.

The end-to-end maximum achievable rate curves in Figure 4.15 are obtained versus the instantaneous SINR condition in the $S \rightarrow D$ link, i.e., versus $\gamma_{SD,i}$. The instantaneous SINR conditions in the $S \rightarrow R$ and $R \rightarrow D$ links are fixed to $\gamma_{SR,i} = 19$ dB and $\gamma_{RD,i} = 30$ dB, respectively. At these SINR conditions, the rate improvement of cooperative-MIMO scheme is maximum as compared to other schemes. The results in this figure show that, cooperative-MIMO scheme can achieve an achievable rate gain of as much as 1.2 b/s/Hz as compared to both w/o relay and cooperative-MISO schemes. The end-to-end maximum achievable rate curves in Figure 4.16 are obtained versus $\gamma_{SD,i}$ with $\gamma_{SR,i} = 30$ dB and $\gamma_{RD,i} = 4$ dB. With these SINR conditions, the rate improvement of cooperative-MISO scheme is maximum as compared to other schemes. The results in this figure show that, cooperative-MISO scheme can achieve end-to-end achievable rate gains over cooperative-MIMO scheme as much as 0.33 b/s/Hz. This is due to the following reason. With cooperative-MISO scheme, one can adjust the transmission rates in the first and second phases based on the channel capacity of each phase. Hence, for very good SINR conditions in the first phase (i.e. for very good $\gamma_{SR,i}$), the transmission rate in the first phase can be very high. This compensates to a certain extent the multiplexing-loss. With cooperative-MIMO scheme on the other hand, one can achieve a higher channel capacity but needs to use the same rate over the two phases. This results in an end-to-end rate determined by one half of the channel capacity offered by cooperative-MIMO scheme.

The end-to-end maximum achievable rate curves in Figure 4.17 are obtained versus $\gamma_{SD,i}$ with $\gamma_{SR,i} = \gamma_{RD,i} = 30$ dB. The results in this figure show that when $\gamma_{SR,i} \gg \gamma_{SD,i}$ and $\gamma_{RD,i} \gg \gamma_{SD,i}$, all the relaying schemes perform the same and the DF-based relaying has better performance. For these channel conditions, conventional relaying should be used as it is the least complex relaying scheme. The analysis shows that, over the region where relaying improves the performance, the cooperative-SIMO scheme can perform as good as the cooperative-MISO scheme, however it cannot outperform it (e.g., as seen in Figures 4.16 and 4.17). The cooperative-MISO scheme can outperform the cooperative-SIMO scheme as far as the SINR condition in the $S \rightarrow R$ link is very good. This is due to the fact that to achieve a cooperative-MISO channel, it is not necessary to use the same modulation mode over the two phases, however, to achieve a cooperative-SIMO channel the same modulation mode needs to be used over the two transmission phases.

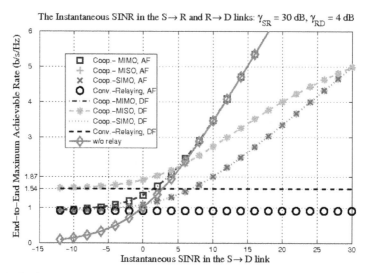

Figure 4.16 The maximum gain of the cooperative-MISO scheme as compared to other schemes.

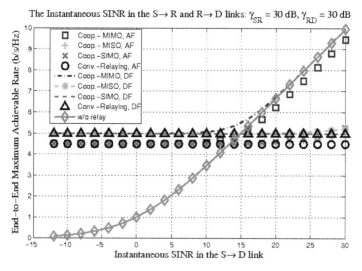

Figure 4.17 When $\gamma_{SR,i} \gg \gamma_{SD,i}$ and $\gamma_{RD,i} \gg \gamma_{SD,i}$, all the relaying schemes perform the same. DF-based relaying has superior performance.

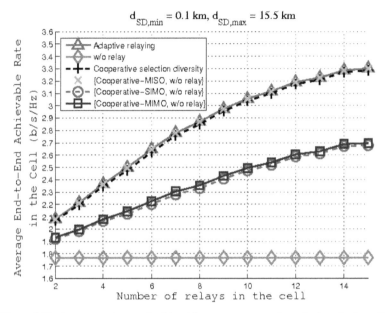

Figure 4.18 Average end-to-end achievable rate versus number of relays in the cell.

In Figure 4.18, the average end-to-end achievable rate in the cell is plotted versus total number of relays in the cell. A single cell with radius 15.5 km where there are a total of 30 users has been simulated. As a practical consideration shown in Chapter 4, the relays are positioned at symmetric positions in the cell at a distance of 10.3 km to the BS. Adaptive relaying refers to the scheme, where the scheme with the best end-to-end achievable rate is used for each instantaneous channel conditions. The PFS developed in Section 4.4 is used with the end-to-end achievable rate equations instead of the throughput obtained from the lookup table. To provide fairness to the users, the PFS window, i.e., T, is set to 100. Relaying is used only if there is an end-to-end rate increase. As the results in Figure 4.18 show, link adaptive transmissions with cooperative-MISO, cooperative selection diversity and adaptive relaying schemes perform similar. This observation agrees with the one presented in Section 4.7. These schemes outperform cooperative-MIMO and cooperative-SIMO schemes thanks to reduction in multiplexing loss via adjusting the transmission rates optimally in each phase. All the schemes outperform w/o relay scheme provided that the best scheme among with relay and w/o relay is selected dynamically based on CSI. Due to the same reasons stated in Section 4.7, the cooperative-MIMO and cooperative-SIMO schemes perform similar in the system level.

4.9 Conclusions and Future Work

In this chapter, a scheduler for implementation in two-hop cellular networks has been developed for low mobility users. This scheduler is able to consider all the instantaneous SINR conditions and the end-to-end throughput in order to schedule the users on the radio resources. It guarantees that the end-to-end performance is always better than that of a w/o relay system and fixed relaying. With this scheduler, the performance of various cooperative diversity schemes has been investigated for two-hop cellular networks with fixed relays. The investigations are provided for the emerging IEEE 802.16j-based wireless relay networks. Interestingly, the investigations show that the cooperative selection diversity is promising when the instantaneous SINR knowledge is available at the transmitter. This is due to the following reasons. From the system-level throughput point of view, the cooperative selection diversity can perform as well as more complex cooperative diversity schemes which require coherent signal combining at the destination. At a given subchannel, it requires transmissions either from the BS or the RS. This reduces the interference in the system as compared to using other cooperative diversity schemes which require simultaneous transmissions from the BS and the RS.

In this chapter, the implementation findings outlined in Chapters 3 and 4 have been complemented by information theoretic performance analysis as well. To this end, the end-to-end maximum achievable rate of various relaying schemes have been analyzed comparatively. The analysis include link adaptive transmissions with both AF and DF-based relaying. The performance comparisons are provided over the SNR region where relaying improves the performance as compared to that of w/o relay transmission. This analysis shows that the DF-based relaying provides significant achievable rate gain over the AF-based relaying for various relaying schemes. The AF-based relaying on the other hand cannot provide significant achievable rate gain as compared to the DF-based relaying. Consequently, the DF-based relaying is promising for the wireless relay networks. The analysis shows that cooperative-MIMO scheme can provide an end-to-end achievable rate gain as compared to other schemes such as cooperative-SIMO, cooperative-MISO, conventional relaying and w/o relay. However, this gain can be achieved when $\gamma_{RD,i} \gg \gamma_{SD,i}$ and $\gamma_{RD,i} \gg \gamma_{SR,i}$. When $\gamma_{SR,i} \gg \gamma_{SD,i}$,

$\gamma_{SR,i} \gg \gamma_{RD,i}$, cooperative-MISO scheme can achieve a higher end-to-end maximum achievable rate as compared to that of other schemes. As far as $\gamma_{SR,i} \gg \gamma_{SD,i}$ and $\gamma_{RD,i} \gg \gamma_{SD,i}$, all the relaying schemes perform the same. The information theoretic conclusions agree with that of a practical system outlined in Chapters 3 and 4.

Future work out of this chapter includes, but is not limited to, the following:

1. The design and evaluation for the users with high speeds where obtaining the instantaneous CSI at the transmitter is not practical.
2. Evaluations for a multi-cell environment.

Acknowledgments

The author would like to thank Euntaek Lim, Persefoni Kyritsi, Simone Frattasi, Megumi Kaneko, Abdulkareem Adinoyi, Hamid Saeedi, Ali Taha Koç, and other involved researchers working at the R&D Center, Samsung Electronics Co. Ltd., Intel Corporation, USA, Center for TeleInFrastruktur (CTIF) and the Department of Systems and Computer Engineering of Carleton University for their valuable comments and/or guidance on this work.

5

Hop Adaptive MAC-PDU Size Optimization for Infrastructure-Based Wireless Relay Networks*

5.1 Introduction

In IEEE 802.16e networks, data is carried in the form of MAC-PDU packets. Once the MAC-PDUs are created at the MAC layer, they are passed to the physical layer for transmission to the destination by using certain bursts. Each burst consists of a certain number of sub-channels. A MAC-PDU in IEEE 802.16e-based networks consists of a fixed length MAC header, variable length payload, optional Fragmentation Subheader (FSH) or Packing Subheader (PSH) and a CRC [21]. The MAC header length is 48 bits [21]. It carries information on the MAC-PDU length, the user ID where the packet is destined to, etc. The payload of a MAC-PDU packet carries the unfragmented or fragmented portion of a MAC-SDU which is delivered to the MAC common part sublayer from the MAC convergence sublayer. Once the MAC-PDUs are created at the MAC layer, they are passed to the PHYsical layer for transmission to the destination. Fragmentation is the process in which a MAC-SDU is fragmented into two or more fragments and transmitted in several MAC-PDUs [21]. This provides flexibility in the MAC-PDU size. When a MAC-PDU contains a fragmented portion of a MAC-SDU, then a FSH is added to each MAC-PDU. The FSH is composed of 2 bits Fragmentation Control (FC), 11 bits Fragment Sequence Number (FSN) for non-ARQ connections, 11 bits Block Sequence Number (BSN) for ARQ-enabled connections and 3 reserved bits. The total length of the FSH is $2 + 11 + 3 = 16$ bits for the extended type. The FC indicates the fragmentation state of the payload, i.e., 00 no fragmentation, 01 last fragment, 10 first fragment, 11 continuing fragment. The FSN indicates the sequence number of the current SDU fragment. The reserved bits are used for the standard to round the length

*This work was conducted at Intel Corporation, Hillsboro, Oregon, USA. © 2008 IEEE. Reprinted with permission from *Proceedings of the IEEE Globecom Conference, 2008.*

of headers to an integer number of bytes. The reserved bits can be used to add additional features to the FSH if needed. This chapter proposes to use these reserved bits to signal to the RS the fragmentation information for the packets it will forward. The details of this operation will be described in Section 5.2. In this chapter, 16 bits CRC is considered.[1] In summary, when a MAC-SDU is fragmented, a total of 48(MAC header) +16(FSH) + 16(CRC) = 80 bits overhead needs to be added per payload, i.e., fragment.

The IEEE 802.16e and IEEE 802.16j standards enable fragmentation of MAC-PDUs. For fragmentation, the current IEEE 802.16j standard does not specify any procedure beyond those specified in IEEE 802.16e or IEEE 802.16-2004 standards [33].

Increasing the packet size (in terms of the total number of bits per packet) increases the PER that each packet reception encounters. Decreasing the packet size decreases the PER. However, this at the same time increases the significance of upper layer overheads carried per packet. Hence, a trade off should be made between the significance of overheads and the PER in choosing a packet size. The MAC-PDU size optimization in multi-hop networks needs to consider the channel condition in each hop. Since channel conditions over different hops are independent and are different, using the same MAC-PDU size over different hops is not efficient. For example, if the MAC-PDU size is decided based on weaker links and kept the same over all hops, then the system will not be able to extract the benefit from stronger links. If the MAC-PDU size is decided based on stronger links and kept the same over all hops, then the weaker links might encounter significant PER. For optimal operation, the MAC-PDU size in each hop should be selected based on the channel condition seen over that hop. This way, the PER experienced at each hop can be minimized. This requires the relays to be able to open up the MAC-PDUs and change the MAC-PDU size before transmitting to the next hop. Therefore, DF-based forwarding is used where the relay decodes, fragments (when necessary), re-encodes and forwards the information bits received from the source.

In this chapter, a method to efficiently implement the MAC-PDU size optimization in OFDMA modulated two-hop wireless relay networks with infrastructure-based relays is proposed. This chapter further proposes a method that reduces the total overhead transmitted in the end-to-end path. The organization of this chapter is as follows. Section 5.2 presents the key concept of the proposal. Section 5.3 presents the system model. In Section 5.4, the

---

[1] Hence, the probability of undetected errors is negligible [44].

effect of MAC-PDU size on the goodput is analyzed. In Section 5.5, a method for the optimization of the MAC-PDU size in two-hop wireless relay networks with infrastructure-based relays is proposed. The performance of this proposal is presented in Sections 5.6 and 5.7 in terms of end-to-end overhead reduction and goodput gain. Conclusions and future works are presented in Section 5.8.

5.2 The Implementation of MAC-PDU Size Optimization with Infrastructure-Based Relays

In this chapter, relaying is chosen only if it can improve the end-to-end goodput as compared to that of direct transmission. If relaying is chosen for a given user, conventional relaying is used and hence the end-to-end burst used for that user contains one relay burst and one access burst used for $RS \rightarrow MS$ transmissions. If relaying is not chosen for a given user, then the *end-to-end burst* used for that user contains one access burst which span the overall duration of the DL frame where data transmissions take place. Each relay and access burst carry unicast data traffic. However, depending on the scheduling and the channel state information of a given user, one user can use more than one burst.

The construction of MAC-PDUs in wireless relay networks can be done in two ways. One way to do it is in an end-to-end basis. The other way is to do it in a hop-by-hop basis. With the end-to-end approach, relays do not change the structure of the MAC-PDUs. They forward the packets as they are. In the hop-by-hop approach, each relay on the end-to-end path can open up MAC-PDUs and can change the fragmentation configuration in it. This allows optimizing the MAC-PDU size in each hop. In Figure 5.1, the end-to-end operation is depicted for a relayed transmission to a given MS in a two-hop DL transmission. Such an operation results in repeating the overheads twice, hence cannot take the advantage of potentially robust $BS \rightarrow RS$ channels with fixed relays. To remedy this problem, a hop adaptive MAC-PDU size optimization is proposed as follows. If relaying is used for a given MS, the relay burst carries the aggregated payload data that is going to be forwarded via the relay in the second hop. Hence, for relayed transmissions, the MAC-PDU size in the first hop will be longer than or equal to that of second hop. This is efficient when the $BS \rightarrow RS$ channels are robust and can carry longer length MAC-PDU packets as compared to that of access links. The relay opens up the MAC-PDU and extracts the MAC header, FSH and payload data.

MAC PDUs Carried in Relay Burst
(BS→RS, first hop)

MAC PDUs Carried in Access Burst
(RS→MS, second hop)

MAC header	FSH	Payload$_1$	CRC
MAC header	FSH	Payload$_2$	CRC
MAC header	FSH	Payload$_3$	CRC
MAC header	FSH	Payload$_4$	CRC

MAC header	FSH	Payload$_1$	CRC
MAC header	FSH	Payload$_2$	CRC
MAC header	FSH	Payload$_3$	CRC
MAC header	FSH	Payload$_4$	CRC

Optimal packet length (in bits)   Optimal packet length (in bits)

Figure 5.1 The conventional non-hop adaptive transmission where the relay does not change the structure of MAC-PDUs. Such operation results in repeating the overheads twice, i.e., the MAC header, FSH and CRC of each packet in the first-hop are repeated in the second-hop.

MAC-PDU carried in the relay burst (BS→RS, first hop)

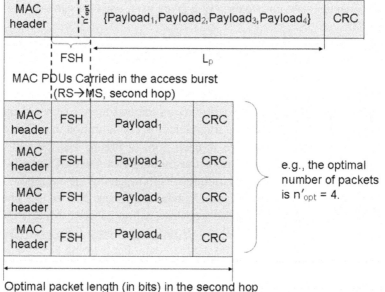

MAC header	n'_{opt}	{Payload$_1$,Payload$_2$,Payload$_3$,Payload$_4$}	CRC

FSH L_p

MAC PDUs Carried in the access burst
(RS→MS, second hop)

MAC header	FSH	Payload$_1$	CRC
MAC header	FSH	Payload$_2$	CRC
MAC header	FSH	Payload$_3$	CRC
MAC header	FSH	Payload$_4$	CRC

e.g., the optimal number of packets is $n'_{opt} = 4$.

Optimal packet length (in bits) in the second hop

Figure 5.2 The proposed hop adaptive MAC-PDU size optimization: The relay changes the structure of MAC-PDUs before forwarding.

The BS informs the RS on the optimal number of packets, i.e., n'_{opt}, for each access burst allocated to transmissions from the relay. The n'_{opt} is proposed to be signalled by using the reserved bits field of the FSH of the MAC-PDU carried in the relay burst. Therefore, such signalling does not incur any additional overhead. The RS fragments the payload data of the MAC-PDU that it has received via the relay burst into n'_{opt} fragments. It modifies the length field of the MAC header and appends it to each fragment. It adds a new FSH for each fragment. It then calculates and adds the CRC to each new packet. Finally, n'_{opt} MAC-PDU packets are created to be transmitted via the relay on the access burst allocated to transmissions from the relay to a given MS. If the optimal number of MAC-PDU packets is more than one for the access burst, then, such hop adaptive approach will save from MAC, CRC and FSH overheads as compared to the end-to-end operation. This procedure is depicted in Figure 5.2.

5.3 System Model

Centralized scheduling and radio resource allocation is done at the BS. A multi-user environment is considered. For each burst, a flat-fading channel is assumed where the fading coefficients remain unchanged over the frequency and time dimensions of the burst. However, the fading level at different bursts can be different based on the frequency selectivity of the channel. Hence, the CSI feedback for each burst is effective. The modulation mode and the packet size to be used for the transmission of MAC-PDU packets are adapted according to the channel condition seen at each burst. Each user feedbacks to the BS the SINR it sees at each burst. Since CSI is available at the BS, AMC is used.

Multiple relays are considered. The performance evaluations can be extended to include the multiple cell scenario. The users do not need to know the existence of the relay. It is assumed that the relays are deployed at strategic positions in the cell such that the first hop $BS \rightarrow RS$ links are much reliable than the second hop $RS \rightarrow MS$ links. The achievement of such deployment has been presented in Chapter 4.

In this chapter, FEC is not considered. The analysis presented in this chapter can be extended to include FEC. When analytical derivations are not available, look-up tables can be prepared to select the MCS and the packet size for a given channel condition. When FEC is included, the throughput of all the schemes will improve [42], hence, the relative performance is expected to remain similar. The considered modulation schemes are BPSK, QPSK, 16-

QAM and 64-QAM. It is assumed that a MAC-PDU packet is discarded at the MS if there is at least one bit received in error. Such error detection can be based on CRC [45]. Each MAC-PDU packet contains a fragment of a given MAC-SDU and hence needs one FSH.

The OFDMA parameters: The OFDMA system parameters considered are based on the IEEE 802.16e standard [45]. The scalable OFDMA parameters with a system bandwidth of 10 MHz and a total of 1024 sub-carriers have been considered. The sub-carrier spacing is 10.94 kHz where the FFT sample rate is 11.2 MHz. The operating carrier frequency is 2.5 GHz. The frames have 5 ms duration. There are a total of 49 OFDMA symbols in each frame. One OFDMA symbol is used for preamble transmission from the BS. One OFDMA symbol duration is used as a guard time between the DL and UL sub-frames. Hence, a total of 47 OFDMA symbols are available for data and control information transmissions in an IEEE 802.16e-based system. In such system, the number of symbols to be used in the DL and UL are given by $47 - n$ OFDMA symbols for DL, and n OFDMA symbols for the UL. The term n is an integer in $12 < n < 21$ [32]. In this chapter, $n = 13$ is chosen. Hence, the DL sub-frame has a total of 34 OFDMA symbols. A total of three OFDMA symbols is used for transmitting FCH, DL and UL-MAP information which are used to signal the radio resource allocation to the users and the relay. In order to enable two hop communication in the TDD mode, one OFDMA symbol duration is needed as a guard time for the relays to switch from the reception mode to the transmission mode. Hence, there remain a total of 30 OFDMA data symbols that can be used for each DL end-to-end burst.

Burst profile: The burst profiles are based on the following configuration. One bin is comprised of nine contiguous sub-carriers. One bin has eight data carrying sub-carriers and one pilot sub-carrier. One slot is composed of two bins in the frequency domain and three OFDMA symbols in the time domain. One sub-channel (i.e., burst) is composed of four bins in the frequency domain and multiple contiguous OFDMA symbols in the time domain. One access burst used for $RS \rightarrow MS$ transmissions is comprised of four bins in the frequency domain and 15 OFDMA symbols in the time domain. One access burst used for w/o relay transmissions is comprised of four bins in the frequency domain and 30 OFDMA symbols in the time domain. The time duration (i.e. total number of OFDMA symbols) of a relay burst can be shorter than or equal to that of access burst and can be up to 15 OFDMA symbols. This is due to the fact that different MCSs can be selected for the relay and access bursts when conventional relaying is used [42]. Consequently, an access burst can carry $60 \leq L \leq 360$ bytes[2] with relay, $120 \leq L \leq 720$ bytes[3] for w/o relay transmissions. For each access and relay burst, the optimal MCS, the optimal MAC-PDU size, and hence the optimal total number of MAC-PDUs are chosen by the BS. These parameters determine the burst profile.

The channel model: The users are pedestrian with a maximum speed of 3 km/h. For such users, the channel model which is referred to as pedestrian-A in [32] has been used. The presented model in [32] has the following power delay profile. The taps are located at $\tau = [0 \ 110 \ 190 \ 410]$ ns. The average power at each tap is given by $P = [0 \ -9.7 \ -19.2 \ -22.8]$ dB. This channel model has a coherence bandwidth of 434.84 kHz ($\lceil 39.75 \rceil$ sub-carriers) where the channel correlation is above 0.9. The channel models presented in [40, 48] can be used to analyze broadband communication with fixed, i.e., non-moving, wireless terminals. Since infrastructure-based relays are considered, wireless channel and path-loss models developed in [40, 48] are used for the $BS \rightarrow RS$ links. The selected model has a power delay profile given by $P = [0 \ -15 \ -20]$ dB and $\tau = [0 \ 0.4 \ 0.9] \ \mu$s. This model corresponds to a coherence bandwidth of 165 sub-carriers where the frequency correlation is above 0.5. It is assumed that the packets in the relay bursts experience a negligible PER with 64-QAM transmissions even if the packet size goes up to 360 bytes in the relay burst. As shown in Chapter 4, this is practical to achieve for LOS conditions in the $BS \rightarrow RS$ link when the infrastructure-based relay is deployed at a strategic position with an appropriate distance to the BS [51].

5.3.1 Terminology

The term j, $j \in \{1, 2, \ldots, J\}$, denotes the burst index in the frequency domain. The term l represents the packet length in bits. $\gamma_{SD,u,j}$ and $\gamma_{RD,u,j}$ represent the instantaneous SINR that a user experiences on burst j of the $BS \rightarrow MS$ and $RS \rightarrow MS$ links, respectively. $L = B \times M$ represents the maximum number of bits that can be carried over the access burst. B represents the total number of data carrying sub-carriers available in the access burst. σ represents the shadowing standard deviation in dB. κ is a pseudo-random number drawn from a normal distribution with mean zero and standard deviation one. k represents the DL frame index in time. The terms

---

[2] I.e., $15 \times 8 \times 4 \times 1 \leq L \leq 15 \times 8 \times 4 \times 6$ bits.
[3] I.e., $30 \times 8 \times 4 \times 1 \leq L \leq 30 \times 8 \times 4 \times 6$ bits.

n'_{opt} and M'_{opt} represent the optimal number of MAC-PDU packets and the optimal MCS both providing the highest goodput when conventional relaying is used. The terms n_{opt} and M_{opt} represent the optimal number of MAC-PDU packets and the optimal MCS both providing the highest goodput when w/o relay transmissions are used.

5.4 The Effect of Adaptive MAC-PDU Size on the Goodput

In this section, the effect of varying MAC-PDU size on the goodput is presented for w/o relay transmissions in the access burst. The end-to-end throughput with a given modulation scheme in a w/o relay system, i.e., $\rho(\gamma_j, l, M)$, is given by [38]

$$\rho(\gamma_j, l, 1) = \left(1 - 0.5\text{erfc}(\sqrt{\gamma_j})\right)^l \text{ b/s/Hz, for BPSK modulation,}$$

$$\rho(\gamma_j, l, 2) = 2\left(1 - 0.5\text{erfc}\left(\sqrt{\frac{\gamma_j}{2}}\right)\right)^l \text{ b/s/Hz, for QPSK modulation}$$

$$\rho(\gamma_j, l, 4) = 4\left(1 - \frac{3}{4}\text{erfc}\left(\sqrt{0.1\gamma_j}\right)\right)^{l/2} \text{ b/s/Hz for 16-QAM modulation}$$

$$\rho(\gamma_j, l, 6) = 6\left(1 - \frac{7}{8}\text{erfc}\left(\sqrt{\frac{1}{42}\gamma_j}\right)\right)^{l/3} \text{ b/s/Hz, for 64-QAM modulation}$$

$$(5.1)$$

for each of the modulation modes, i.e. BPSK ($M = 1$), QPSK ($M = 2$), 16-QAM ($M = 4$) and 64-QAM ($M = 6$), respectively. The term γ_j represents the short-term average SINR condition observed at burst j. In these equations, the effect of overhead on the net throughput delivered to the upper layers is not considered. When addition of the headers and the CRC overhead is not considered, the decreasing MAC-PDU payload size will increase the throughput. This is due to the fact that, the decreasing packet size decreases the packet error rate for a given SINR. With decreasing MAC-PDU size, the significance of header and CRC overheads increases. This in turn decreases the goodput. Hence, a trade off should be made between increased reliability and increased overhead by taking into account of goodput. Such a trade off can be seen in Figure 5.3.

In the figure, the effect of varying MAC-PDU size is analyzed according to the following set-up. A MAC-PDU is comprised of a total of $l \in$

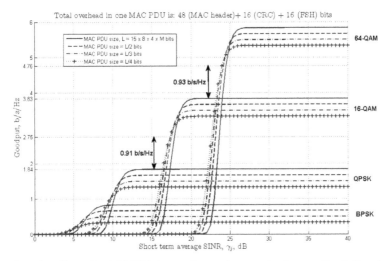

Figure 5.3 The effect of MAC-PDU size on the goodput received over a flat fading channel. The MAC, FSH and CRC overhead are considered.

$\{L, L/2, L/3, L/4\}$ bits including a total overhead of 80 bits per MAC-PDU. For a MAC-PDU size of l bits, the goodput in terms of

$$\rho(\gamma_j, l, 6) \times \frac{l - 80}{l}, \quad \rho(\gamma_j, l, 4) \times \frac{l - 80}{l},$$

$$\rho(\gamma_j, l, 2) \times \frac{l - 80}{l} \quad \text{and} \quad \rho(\gamma_j, l, 1) \times \frac{l - 80}{l}$$

are plotted for a flat fading channel with varying short-term average SINR γ_j. As the results in Figure 5.3 show, the optimal MAC-PDU size depends on the channel condition. Using the optimal packet size for each channel condition results in significant goodput gain. When the channel can enable error free transmissions for a modulation mode say M, the optimal solution with AMC is to increase the MAC-PDU size until the whole packet fits into one burst. When PER is non-negligible, then the optimal solution is not only to choose the most efficient MCS but also to choose the most efficient packet size. When PER is non-negligible, reducing the packet size can increase the goodput even if there is an increasing significance of overhead.

5.5 Optimization of the MAC-PDU Size in Two Hop Wireless Relay Networks with Infrastructure-Based Relays

5.5.1 Method for Hop Adaptive MAC-PDU Size Optimization

Based on CSI, the BS decides on the optimal number of MAC-PDU packets (i.e., the optimal size of each packet) and the MCS to be used in an access burst. The MCS and the packet size are optimized jointly based on analytical derivations. Such optimization is done both for with and w/o relay cases according to

$$\varphi_{u,j}^{\text{with}-\text{relay}}(n'_{\text{opt}}, M'_{\text{opt}})$$

$$= \max_{n \in \mathbb{N}} \left\{ \max_{M \in \{1,2,4,6\}} \left\{ \frac{L_p \rho(\gamma_{SR,j}, L_1, 6) \rho(\gamma_{RD,u,j}, L/n, M)}{(ML_1 + 6L)} \right\} \right\}, \quad (5.2)$$

$$\varphi_{u,j}^{\text{w/o}-\text{relay}}(n_{\text{opt}}, M_{\text{opt}})$$

$$= \max_{n \in \mathbb{N}} \left\{ \max_{M \in \{1,2,4,6\}} \left\{ \frac{L_p}{L} \rho \left(\gamma_{SD,u,j}, \frac{L}{n}, M \right) \right\} \right\}. \quad (5.3)$$

The term n represents the number of MAC-PDU packets to be transmitted in the access burst and it is an integer such that L/n is an integer. Hence, for a total of n packets to be transmitted over the access burst, the total number of bits included in each MAC-PDU is given by L/n. The term $L_p = L - n80$ represents the total payload bits transmitted in the access and relay bursts. With hop adaptive MAC-PDU size optimization proposed in this chapter, the total number of bits transmitted via the relay burst in one MAC-PDU is given by

$$L_1 = \left\lceil \frac{(L_p + 80)}{6} \right\rceil \times 6.$$

The number 80 represents the total overhead in one MAC-PDU packet. The term, $\varphi_{u,j}^{\text{with}-\text{relay}}(n'_{\text{opt}}, M'_{\text{opt}})$ represents the optimal goodput that user u can obtain at the end-to-end burst j via the conventional relay transmissions. The term $\varphi_{u,j}^{\text{w/o}-\text{relay}}(n_{\text{opt}}, M_{\text{opt}})$ represents the optimal goodput that user u can obtain at the end-to-end burst j via w/o relay transmissions.

For the end-to-end burst, the BS finally decides whether to use the relay or not via

$$\varphi_{\text{opt},u,j}[k] = \max\{\varphi_{u,j}^{\text{with}-\text{relay}}(n'_{\text{opt}}, M'_{\text{opt}}), \varphi_{u,j}^{\text{w/o}-\text{relay}}(n_{\text{opt}}, M_{\text{opt}})\}. \quad (5.4)$$

If $\varphi_{\text{opt},u,j}[k] = \varphi_{u,j}^{\text{with}-\text{relay}}(n'_{\text{opt}}, M'_{\text{opt}})$, then conventional relaying is used in the end-to-end burst which is comprised of one relay burst and one access burst. If $\varphi_{\text{opt},u,j}[k] = \varphi_{u,j}^{\text{w/o}-\text{relay}}(n_{\text{opt}}, M_{\text{opt}})$, then w/o relay transmission is used in the end-to-end burst.

Once the optimal MCS, optimal packet size and optimal transmission mode with or w/o relay are chosen for a given user u, then the BS schedules the users on bursts based on a modified PFS algorithm given by

$$\hat{u} = \text{argmax} \left\{ \frac{\varphi_{\text{opt},u,j}[k]}{\overline{\varphi}_u[k-1]} \right\}. \tag{5.5}$$

If $\hat{u} = u$, then the transmissions for user u are scheduled on burst j. The term $\overline{\varphi}_u[k-1]$ represents the past average goodput of user u at the previous DL frame $k-1$. Once the users are scheduled, the past average goodput for each user is updated using a low pass filter with a time constant of T slots. This update is done according to

$$\overline{\varphi}_u[k] = \frac{(T-1)\overline{\varphi}_u[k-1] + \sum_{j=1}^{J}(c_{u,j}\varphi_{\text{opt},u,j}[k])}{T}. \tag{5.6}$$

The term $c_{u,j}$ is equal to one if transmissions for user u is scheduled on burst j, otherwise it is equal to zero. This scheduler provides both cooperative diversity and multi-user diversity. The cooperative diversity is achieved by dynamically selecting the best link among with or w/o relay. For a given user, it guarantees that the end-to-end throughput will always be greater than or equal to that of w/o relay. The time constant T adjusts the level of fairness of the scheduler. The time constant T is set to $T = 100$ in order to provide a fair channel allocation to the users (see Chapter 4).

5.6 Link Level Performance Evaluation

This section provides link level performance evaluation of the hop adaptive MAC-PDU optimization for transmissions over a given end-to-end burst j. In Figure 5.4, the gain in goodput via the use of hop adaptive optimal MAC-PDU size selection is presented for conventional relaying. $\gamma_{RD,j}$ represents the short-term average SINR condition experienced in the access burst. It is determined by the fading and shadowing experienced at the access burst and can be different from frame to frame. The solid line corresponds to the case where, the packet size in each hop is the same and it is simply determined by the chosen optimal MCS, i.e., $l = B \times M'_{\text{opt}}$. The dashed line corresponds

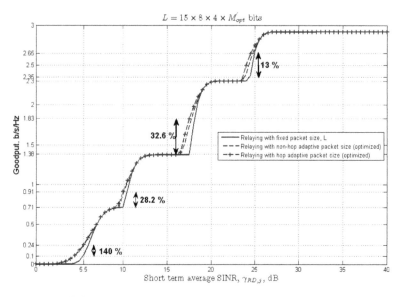

Figure 5.4 The gain in end-to-end goodput when the packet size is optimized as well as the MCS.

to the case where, the packet size is optimized for the access burst and it is kept the same in the relay burst. This is referred to as non-hop adaptive packet size optimization. The dashed line with + signs corresponds to the case where the packet size is optimized in each hop based on the hop adaptive optimal MAC-PDU size selection procedure described herein. As the curves show, the packet size optimization (either hop adaptive or non-hop adaptive) brings significant goodput improvement as compared to without packet size optimization. For example, the goodput gains are 140, 28.2, 32.6 and 13% for SNRs 5.5, 10, 17.5 and 24.5 dB, respectively. These gains vary according to SNR condition seen over the burst. When the channel condition can enable transmission even with longest packet size that can fit to the burst, the results show that packet size optimization is not needed for those channel conditions. The hop-adaptive packet size optimization does not bring significant goodput improvement as compared to non-hop adaptive packet size optimization. However, hop-adaptive packet size optimization does bring significant gains in overhead (e.g., as shown in Figures 5.1, 5.2 and 5.5). Let Θ represent the overhead gain in percent achieved with the hop adaptive MAC-PDU size

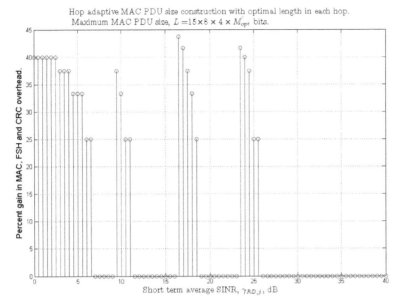

Figure 5.5 The gain in total overhead when the MAC-PDU size is optimized for each hop. The gain is calculated as compared to the non-hop adaptive packet size optimization.

optimization proposed herein. This gain is given by

$$\Theta = \frac{2n'_{\text{opt}}80 - (n'_{\text{opt}} + 1)80}{2n'_{\text{opt}}80} \times 100.$$

For each short-term average SINR in the access burst, overhead gain Θ is depicted in Figure 5.5. In Figure 5.6, n'_{opt} is plotted versus $\gamma_{RD,j}$.

As the results show, when $n'_{\text{opt}} > 1$, the gain in overheads is greater than or equal to 25% and can go up to 43.75%. n'_{opt} goes up to 8. Hence, only 3 bits are necessary to inform the relay on n'_{opt}. In this chapter, the reserved bits of FSH are proposed to be used to inform the relay on n'_{opt}. This adds a new feature for the relay support in 802.16-based networks.

5.7 System Level Performance Evaluation

5.7.1 The Simulation Setup

In this section, the simulation setup is presented. The relays are deployed symmetrically in the cell. When necessary, an MS receives transmissions of

Optimal number of packets in the burst with optimal AMC mode at each channel condition.

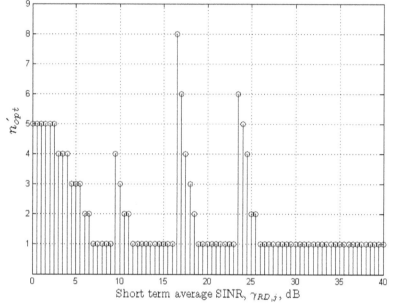

Figure 5.6 The n'_{opt} versus short-time average SINR of the access burst.

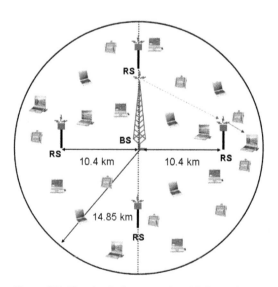

Figure 5.7 The simulation scenario with four relays.

the BS via the relay with the closest distance to it. This scenario is depicted in Figure 5.7 for a total of four relays. A total of 23 users and 23 end-to-end bursts are simulated. The RS and each MS have a noise figure of 7 and 9 dB, respectively.

The path-loss models: The distance between BS and a given RS is $d_{SR} = 10.4$ km. With the system model considered in this book, this results in efficient deployment of the relays as presented in Chapter 4. At this distance, the relays can decode the MAC-PDUs received from the BS with negligible block error rate with the path-loss model presented in the following. The $BS \rightarrow RS$ link has a path-loss exponent of $n_{SR} = 3$ and a LOS K factor of $K = 10$. The $RS \rightarrow MS$ and $BS \rightarrow MS$ links are NLOS and have a path loss exponent of $n_{MS} = 3.5$. The carrier frequency is $f_c = 2.5$ GHz. The wavelength λ is given by $\lambda = 3 \times 10^8/f_c$. The path-loss in the $BS \rightarrow RS$ link is calculated according to [40]:

$$\mathrm{PL_{BS,RS}} = A + 10 n_{SR} \log_{10}\left(\frac{d_{SR}}{do}\right)$$

$$+ 6 \log_{10}\left(\frac{f_c}{10^6 2000}\right) - 20 \log_{10}(h_{RS}/2) \text{ dB.} \quad (5.7)$$

h_{RS} is the height of the RS antenna from the ground level and it is assumed to be 10 m. The term A is given by $A = 20 \log_{10}(4\pi do/\lambda)$. The reference distance d_o is 100 m. The term $6 \log_{10}(f_c/10^6 2000)$ represents the frequency correction factor which reflects higher path loss at higher frequencies [63,64]. The term $20 \log_{10}(h_{RS}/2)$ represents the amount of reduced path loss thanks to higher antenna heights [65]. The RS is assumed to have a $\Phi_{RS} = 15$ dBi of receive antenna gain. Hence, the received signal power from the BS at the RS is given by

$$P_{Rx,BS,RS} = 10^{((10 \log_{10}(P_{BS}) - \mathrm{PL_{BS,RS}} + \Phi_{RS})/10)} \text{ W,}$$

where the EIRP from the BS is set to $10 \log_{10}(P_{BS}) + 30 = 57.3$ dBm [3].

For a BS, MS distance of $d_{BS,MS}$ m, the path-loss in a $BS \rightarrow MS$ link and the received signal power from the BS at the MS are given by

$$PL_{BS,MS} = A + 10 n_{MS} \log_{10}\left(\frac{d_{BS,MS}}{do}\right) + 6 \log_{10}\left(\frac{f_c/10^6}{2000}\right) + \sigma \kappa \text{ dB,} \quad (5.8)$$

$$P_{Rx,BS,MS} = 10^{((10 \log_{10}(P_{BS}) - PL_{BS,MS} + \Phi_{MS})/10)} \text{ W,} \quad (5.9)$$

respectively. The receive antenna gain at an MS is assumed to be $\Phi_{MS} = -3$ dBi. The shadowing standard deviation σ is 8 dB.

For an RS, MS distance of $d_{RS,MS}$ m, the path-loss in an RS $\rightarrow$ MS link and the received signal power from the RS at the MS are given by

$$PL_{RS,MS} = A + 10n_{MS} \log_{10} \left(\frac{d_{RS,MS}}{do} \right)$$

$$+ 6 \log_{10} \left(\frac{f_c/10^6}{2000} \right) + \sigma\kappa \text{ dB}, \tag{5.10}$$

$$P_{Rx,RS,MS} = 10^{(10 \log_{10}(P_{RS}) - PL_{RS,MS} + \Phi_{MS})/10} \text{ W}, \tag{5.11}$$

respectively. The EIRP from the RS is set to $10 \log_{10}(P_{RS}) + 30 = 47.3$ dBm.

5.7.2 Performance Evaluation

With this simulation setup, the system level simulations are done and presented in this section. The transmissions for the users are scheduled on each burst based on the PFS presented in Equations (5.5) and (5.6). The system level simulations show the results presented in Table 5.1. The average goodput results are obtained when the goodput is greater than or equal to 0.001 b/s/Hz. The first column in the table represents the total number of relays in the cell. The second column represents the average gain in header and CRC overheads in the relayed links. The third column represents the average gain in the end-to-end goodput via MAC-PDU size optimization. As the results show, there is a significant performance improvement in the system level. For example, when there are a total of three relays in the cell, MAC-PDU size optimization brings an end-to-end average goodput gain of 13%. On top of such gain, our proposal brings an average overhead gain of 13.52% in the relayed links. The system level end-to-end goodput gain reduces with increasing number of relays. This is due to the fact that, when there are more relays in the system, a user has more chance to obtain a better channel condition. This in turn reduces the goodput gain coming from MAC-PDU size optimization. Even with increasing number of relays in the cell, the proposed hop adaptive MAC-PDU size optimization still brings significant gain in header and CRC overheads transmitted in the end-to-end path.

5.8 Conclusions and Future Work

In this chapter, the inefficiency of having the same MAC-PDU size over all the links constituting a multi-hop wireless communication system is investig-

Table 5.1 The average system level gains versus number of relay stations.

Number of relays	Average overhead gain (%)	Average goodput gain (%)
2	13.7	17.5
3	13.52	13
4	12.75	7.25
6	12.1	6.2

ated. To remedy this inefficiency and to provide an end-to-end optimization, a hop adaptive MAC-PDU size optimization is proposed. This proposal is analyzed for two-hop wireless relay networks with infrastructure-based relays deployed at strategic locations in the cell. The performance metrics are overhead reduction and goodput gain provided by the hop adaptive MAC-PDU size optimization. When there are a total of three infrastructure-based relays, a single BS, and 23 low mobility users in a Mobile-WiMAX network with relay support, the system level evaluations show that the average gain in goodput is 13% and average gains in the header and CRC overheads in relayed links is 13.54%. The investigations show that the average goodput gain via MAC-PDU size optimization reduces with increasing number of relays in the cell. On the other hand, the proposal brings significant reductions in the header and CRC overheads even with increasing number of relays in the cell.

Future work out of this work includes the following:

1. Design and analysis for high mobility users where the channel varies more often.
2. Inclusion of FEC.
3. Design and evaluation for a multi-cell environment.
4. Design and evaluation with reduced and/or imperfect CSI.
5. Design and analysis with ARQ and HARQ.

Acknowledgments

The author would like to thank Liuyang (Lily) Yang, Jaroslaw (Jerry) Sydir, Minnie Ho, Professor Halim Yanikomeroglu and other researchers working at Intel Corporation for their valuable comments, support and attention on this work.

6

Implementation Issues for OFDM(A)-Based Wireless Relay Networks*

This chapter presents several implementation issues for OFDMA-based multi-hop cellular networks which need inclusion of relay terminals. The first issue presented is synchronization. It is shown that synchronization is not problematic for infrastructure-based relaying, where the relays are deployed by a system operator at strategic positions in the cell. Another implementation issue that is presented is related to the hardware implementation aspects. The hardware resource usage analysis shows that cooperative diversity schemes which require coherent signal combining at the MS increase hardware resource usage and power/energy consumption at the MSs. Furthermore, the channel estimation errors at the MS result in an effective decrease in the post-processing SINR achieved via coherent signal combining, e.g., MRC [66]. The reliability of channel estimation at the MSs depends on the density and power of pilot tones and the algorithm used for channel estimation.

6.1 Synchronization Issues for OFDM(A)-Based Wireless Relay Networks

The frame structure developed in Chapter 3 and the cooperative transmit diversity enable simultaneous transmissions from the BS and the RS terminals. This necessitates investigations on various synchronization issues that will

---

*The work presented in this chapter was sponsored by Aalborg University and Telecommunication R&D Center-Samsung Electronics Co. Ltd., Suwon, Republic of Korea. A part from this work has been published in [1] and [2]. © 2008 IEEE. Reprinted with permission from *Proceedings of the 21st Canadian Conference on Electrical and Computer Engineering, May 2008*. © 2007 IEEE. Reprinted with permission from *IEEE Communications Magazine, Technologies in Multi-Hop Cellular Networks*, September 2007.

arise by enabling simultaneous transmissions from the BS and the RS terminals. In this section, such synchronization issues for the DL transmissions in OFDM(A)-based two-hop cellular networks are investigated.

6.1.1 Introduction in Time Offset Problem

In conventional multi-antenna schemes where the transmit and receive antennas are co-located at a given terminal, the signals that are simultaneously transmitted by the transmitter antennas arrive at the receiver simultaneously since they are transmitted from the same terminal. However, in cooperative transmission schemes where the signals transmitted by the BS and the RS are supposed to be received simultaneously at the receiving terminal, the receiver (i.e., the MS) might experience a time offset between the OFDM(A) symbols that are received from the BS and the RS. This time offset is caused by the geographical separation of the transmitters and the timing imperfection at the relay. Let Δ represent the time offset in seconds. This time offset issue is visualized in Figure 6.1. In the figure, Δ_{RD} and Δ_{SD} represent the propagation delay in the $R \rightarrow D$ and $S \rightarrow D$ links, respectively. In order to have the most robustness to this time offset problem, the MS should align its FFT window with the earliest arriving link as depicted in the figure. Such alignment is considered in this chapter. The term δ in seconds represents the time-delay introduced at the RS before it starts its transmissions. A positive time offset indicates that the signals that are transmitted from the RS arrive first. A negative time offset indicates that the signals that are transmitted from the BS arrive first.

For OFDM-based networks, the time offset does not cause throughput degradation if $|\Delta| + 5\sigma_{rms} < \tau_{CP}$, where τ_{CP} represents the cyclic prefix (CP) duration and σ_{rms} represents the rms delay spread of the channel between the transmitters and the receiver [49]. If $|\Delta| + 5\sigma_{rms} > \tau_{CP}$ then ISI occurs with power directly proportional to:

- the average SINR condition in the RS $\rightarrow$ MS link if the OFDM(A) signals received from the RS arrive after the OFDM(A) signals received from the BS,
- the average SINR condition in the BS $\rightarrow$ MS link if the OFDM(A) signals received from the BS arrive after the OFDM(A) signals received from the RS,
- the total duration of the ISI relative to the symbol duration, i.e., $(|\Delta| + 5\sigma_{rms} - \tau_{CP})/T_{symb}$,

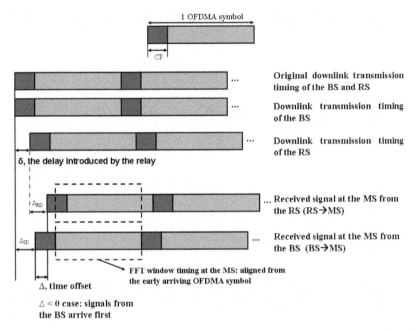

Figure 6.1 The time offset problem and the alignment of the FFT window at the MS in order to achieve most resilience against the time offset problem.

where it is assumed that the FFT window is aligned with the earliest arriving signal. T_{symb} represents the OFDM(A) symbol duration. If only one terminal is allowed to transmit in the second phase (either the BS or the RS), then the time offset problem does not occur, however, in this case the radio resources are not used efficiently as shown in Chapter 3.

In the following, the effects of the time offset is analyzed at various positions in the cell. The same system set-up presented in Chapter 4 has been considered with the following exceptions. The analysis is presented for a single RS. The analysis can be extended to the multiple RS case. The BS is at the center of a cell with radius 9.8 km. The RS is placed at a distance of 4.3 km from the BS, i.e., $d_{SR} = 4.3$ km. With the current system setup, the reasons for this placement is the following. First of all, at this position, it is possible to detect 64-QAM symbols with negligible error rate at the RS. Hence, the benefits of relaying can be exploited efficiently. Furthermore, since the RS is closer to the BS, the time offset may cause significant ISI as the average SINRs in the $S \rightarrow D$ and $R \rightarrow D$ links will be comparable to each other. Hence, this will enable analysis on how worse such ISI might

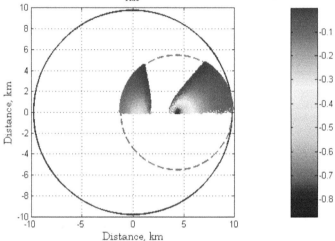

$P_{BS} - P_{RS} = 10$ dB, $R_{BS} = 9.8$ km, $R_{RS} = 5.5$ km, $d_{SR} = 4.3$ km

SINR(dB)- SNR(dB) per position where $\sigma_{rms} = 22$ samples, CP = 256 samples, $\delta = 80$ samples

Figure 6.2 The SINR (dB) – SNR (dB) per position (over the region where time offset causes ISI) within the coverage area of the BS and the RS where the RS transmits its signals with a delay of 80 samples as compared to the instant where the BS starts its transmissions. A full colour version is available on http://www.researchwebshelf.com/DocuDetails.php?SrlNo=46.

affect the performance. The time offset effect is analyzed within a circular area around the relay, which has a radius of 5.5 km.

The investigation presented in Figure 6.2 is obtained over the region where the time offset causes a non-zero ISI. Over the regions presented in white, the time offset does not cause ISI. It can be observed in Figure 6.2 that, when the RS introduces a delay, i.e., δ samples, to its transmissions, the difference between SINR (including the interference caused by ISI, i.e., time offset) and SNR (when the interference caused by the time offset problem is not considered) is less than 1 dB. To achieve this value, the MS should align its FFT window from the earliest arriving link as depicted in Figure 6.1. Consequently, the gain in throughput comes at the cost of a loss of as much as 1 dB in SNR only at very close distances to the RS. Over this region, conventional relaying can be used as the SNR of the RS $\rightarrow$ MS links are very strong as compared to BS $\rightarrow$ MS links. Hence, a loss of 1 dB in SNR will not cause significant degradation. The results presented in Figure 6.2 are obtained for an rms delay spread of 1 μs which corresponds to 22 samples with a sampling frequency of 22.4 MHz of IEEE 802.16e standard. In a practical system, the rms delay spread can be less than 1 μs [41]. In this case, the degradation due to time offset will be less than the one depicted in Figure 6.2. Furthermore, if the relay is positioned at a further distance to the BS, the SNR difference in the BS $\rightarrow$ MS and RS $\rightarrow$ MS links increases (e.g., as shown in Chapter 4). In such a case, the degradation will be less than the one depicted in Figure 6.2. In summary, the time offset problem does not cause significant degradation on the system throughput provided that the RS introduces an appropriate delay in its transmissions. The amount of the necessary delay depends on the position of the RS in the cell and the transmit power difference between the BS and the RS. The time offset Δ is defined as

$$\Delta = \Delta_{SD} - (\Delta_{RD} + \delta). \tag{6.1}$$

6.1.2 Introduction in Frequency Offset Problem

The Carrier Frequency Offset (CFO) problem in OFDM-based wireless networks causes Inter Carrier Interference (ICI) and a common phase error [67]. In conventional multi-antenna schemes, the DL signals received at a given MS suffer from a single CFO, i.e., the CFO caused by the Local Oscillator (LO) mismatch between the BS and the MS. For the relaying schemes where simultaneous transmissions from the BS and the RS are needed, the MS suffers from two distinct CFO effects simultaneously.[1] One of them results from the CFO between the BS and MS, and the other one results from the CFO between the RS and MS. It is not possible for the MS to compensate the CFO for both links at the same time. For infrastructure-based relay terminals, the CFO problem can be solved if the RS estimates its carrier frequency mismatch with the carrier frequency of the BS and compensates for this offset before its transmissions.[2] This way, the MS sees only one CFO caused by the LO mismatch between its own LO and the LO of the BS. Such offset can be compensated at the MS by using existing CFO compensation algorithms used in single-hop cellular networks.

The following synchronization problems are identified for the UL transmissions in the OFDM(A)-based two-hop cellular networks. Assuming that the infrastructure-based relay terminals can align their carrier frequency with

---

[1] It is assumed that the effects of the CFO in the BS $\rightarrow$ RS link in DL and in the RS $\rightarrow$ BS link in UL transmissions are compensated perfectly. This can be achieved with infrastructure-based relays and existing CFO compensation algorithms developed for single-hop cellular networks.

[2] This estimation can be done at a high accuracy since line-of-sight and high SNR conditions in the BS $\rightarrow$ RS links can be achieved with infrastructure-based relay terminals deployed at strategic positions in the cell.

that of the BS, then the CFO problem in the UL of two-hop cellular networks converges to the problems in the single-hop cellular networks. Regarding the time offset problem in the UL, the relays can be treated as the users. Hence, the existing solutions for the time and CFO problems in the UL of single-hop cellular networks can be used for the synchronization problems in the UL of two-hop cellular networks.

In the following section, the analytical model of the CFO effects in DL is developed that justifies the conclusions presented herein.

6.1.3 The Analytical Model of the Time and Frequency Offset Problem in OFDM(A)-Based Wireless Relay Networks

In this section, baseband model of the time and frequency offset problem is presented for OFDM-based two-hop cellular networks with a single relay.

In order to present the analytical model of the time and frequency offsets, the following parameters are defined. The terms ω_{BS}, ω_{RS} and ω_{MS} represent the carrier frequency of the BS, RS and MS, respectively. The terms $\tilde{h}_{SD}(t)$ and $\tilde{h}_{RD}(t)$ represent the baseband time domain channel in the BS $\rightarrow$ MS and RS $\rightarrow$ MS links, respectively. The terms $\tilde{s}_{BS}(t)$ and $\tilde{s}_{RS}(t-\delta)$ represent the transmitted baseband signals by the BS and RS, respectively. With these definitions, the received baseband signal at the MS in the second phase is given by

$$
\tilde{r}_{MS}(t) = \left(\tilde{s}_{BS}(t) * \tilde{h}_{SD}(t - \Delta_{SD})\right)e^{j(\omega_{BS}-\omega_{MS})t}
$$

$$
+ \left(\tilde{s}_{RS}(t) * \tilde{h}_{RD}(t - \Delta_{RD} - \delta)\right)e^{j(\omega_{RS}-\omega_{MS})t} + n_{MS}(t), \quad (6.2)
$$

where $n_{MS}(t)$ represents the AWGN noise at the MS terminal. As seen in the above equation, there are two simultaneous CFO that affects the performance. Due to this fact, the perfect compensation of the CFO can be done only for one link but not for both. If the CFO is compensated for the RS $\rightarrow$ MS link, then the compensated signal at the MS can be written as

$$
\tilde{r}_{MS}(t)e^{-j(\omega_{RS}-\omega_{MS})t} = \left(\tilde{s}_{BS}(t) * \tilde{h}_{SD}(t - \Delta_{SD})\right)e^{j(\omega_{BS}-\omega_{RS})t}
$$

$$
+ \left(\tilde{s}_{RS}(t) * \tilde{h}_{RD}(t - \Delta_{RD} - \delta)\right) + n_{MS}(t). \quad (6.3)
$$

In this case, the power of the ICI caused by the residual CFO appearing on the signal received from the BS $\rightarrow$ MS link will be proportional to the received power from the BS $\rightarrow$ MS link. If the CFO is compensated for the BS $\rightarrow$ MS link instead of RS $\rightarrow$ MS link, then the compensated signal at the MS can be written as

$$\tilde{r}_{\text{MS}}(t)e^{-j(\omega_{\text{BS}}-\omega_{\text{MS}})t} = \left(\tilde{s}_{\text{BS}}(t) * \tilde{h}_{SD}(t - \Delta_{SD})\right)$$
$$+ \left(\tilde{s}_{\text{RS}}(t) * \tilde{h}_{RD}(t - \Delta_{RD} - \delta)\right)e^{j(\omega_{\text{RS}}-\omega_{\text{BS}})t} + n_{\text{MS}}(t). \quad (6.4)$$

In this case, the power of the ICI caused by the residual CFO appearing on the signal received from the RS $\rightarrow$ MS link will be proportional to the received power from the RS $\rightarrow$ MS link. Over the region where relaying improves performance, the received signal power from the RS is higher than that of the signal received from the BS. In this case, it is advantageous to compensate the CFO in the RS $\rightarrow$ MS link instead of BS $\rightarrow$ MS link. If the CFO is compensated for the RS $\rightarrow$ MS link, then the residual CFO will appear on the signal received from the BS $\rightarrow$ MS link which has relatively lower receive power than that of RS $\rightarrow$ MS link over the region where relaying improves performance. This will relieve to a certain extent, the ICI effects. Furthermore, as seen in Equations (6.3) and (6.4), when the CFO is compensated for either one of the RS $\rightarrow$ MS or BS $\rightarrow$ MS links, the residual CFO becomes equal to $|\omega_{\text{BS}} - \omega_{\text{RS}}|$. This is an important observation. With infrastructure-based relays, the CFO of $|\omega_{\text{BS}} - \omega_{\text{RS}}|$ is expected to be less than $|\omega_{\text{BS}} - \omega_{\text{MS}}|$. This is mainly due to the fact that the hardware structure of an infrastructure-based relay can be more robust than that of an MS. The infrastructure-based relay can estimate the $|\omega_{\text{BS}} - \omega_{\text{RS}}|$ with a high accuracy. If a given RS compensates for this CFO prior to its transmissions, then the CFO problem in two-hop relay networks resorts to the CFO problem seen in single-hop networks. This is practical to achieve with infrastructure-based relay terminals.

6.1.4 The Time and Frequency Offset Conditions Resulting in Significant Degradation

In this section, the time and frequency offset conditions resulting in significant degradation are presented with system level simulations. The investigations are presented for various cooperative diversity schemes presented in Chapter 3. Performance with both AF and DF-based relaying is presented.

6.1.4.1 The System Model Considered
A single relay is used for a given user. The cooperative-MIMO and cooperative-MISO schemes are used to achieve the cooperative transmit diversity-1 and cooperative transmit diversity-2 schemes, respectively.

Table 6.1 The simulation parameters [3].

Parameter	Value
Carrier frequency, f_c	2.5 GHz
Sampling frequency, B	22.857 MHz
Total number of sub-carriers, N	360 sub-carriers are simulated
Sub-carrier spacing, Δf	22.857 MHz/360 = 63.5 kHz
CP samples, N_g	256
OFDM symbol duration, T_s	$\frac{N+CP}{B} = 26.95\ \mu\text{sec}$
$S \rightarrow R$ link	LOS, SUI-I: mostly flat terrain
(suburban area) [40, 48]	with light tree densities
	multi-path power delay profile:
	$p = [0 \quad -15 \quad -20]$ dB,
	$\tau = [0 \quad 0.4 \quad 0.9]\ \mu\text{s}$,
	K factor = $[4 \quad 0 \quad 0]$
	$\text{SNR}_{SR} = 48.5$ dB
$S \rightarrow D$ and $R \rightarrow D$ links [41]	NLOS, multi-path power delay profile:
	$p = -1 \times [0 \quad 6.2 \quad 2.6 \quad 10.4 \quad 16.47 \quad 22.2]$ dB,
	$\tau = [0 \quad 0.06 \quad 0.14 \quad 0.4 \quad 1.38 \quad 2.83]\ \mu\text{s}$,
	RMS delay spread = $0.23\ \mu\text{s}$
	For frequency correlation above 0.5,
	coherence bandwidth = 0.87 MHz

Cooperative-SIMO scheme is used to provide cooperative receive diversity. In order to analyze the effects in conventional systems, it is assumed that one type of cooperative diversity scheme is used with fixed relaying. Hence, the time and carrier frequency offset issues become specific to cooperative transmit diversity-1 and 2. It is assumed that the CFO effects in BS $\rightarrow$ RS link, w/o relay, conventional relaying and cooperative receive diversity schemes can be compensated perfectly. AMC is used based on link adaptation and selection method developed in Chapter 3. However, for conventional relaying and cooperative transmit diversity-2, two phases with equal duration (i.e., the same MCS is used in the two hops) has been considered with DF-based relaying. The modulation levels considered are BPSK, QPSK, 16-QAM and 64-QAM. FEC is not implemented. The simulation parameters that are considered are given in Table 6.1. The $S \rightarrow R$ link has an SNR of 48.5 dB which allows error free detection of 64-QAM symbols. Hence, the decoding errors at the relay are negligible.

6.1.4.2 Effects of Residual CFO On the Performance
In this section, the throughput degradation caused by the residual CFO between the BS and the RS is presented. For the second-phase of coopera-

tive transmit diversity-1 and cooperative transmit diversity-2 schemes, the effect of a residual CFO in either one of the RS → MS or BS → MS links is analyzed while assuming that the CFO in one of the links is compensated perfectly. Hence, the residual CFO is equal to $|\omega_{BS} - \omega_{RS}|$ as given in Equations (6.3) and (6.4).

In Figures 6.3, 6.4, 6.5 and 6.6, the relative gain in average spectral efficiency in % is presented versus the residual carrier frequency offset in terms of

$$\frac{100 \times f_{\text{err}}}{\Delta f} = \frac{100 \times |\omega_{BS} - \omega_{RS}|}{2\pi \, \Delta f}.$$

It is assumed that the time offset is within the limits where it does not cause any degradation. For all these figures, the average SNR value for the BS → MS link is set to $SNR_{SD} = 4.21$ dB. Among the figures, only the average SNR in the RS → MS link is changing from 8 dB to 9.5 dB. This is a practical range of SNRs where relaying will be beneficial over w/o relay transmissions. In Figure 6.3 the gain in average spectral efficiency is presented for the case where the CFO is compensated for the BS → MS link perfectly as presented in Equation (6.4). In Figure 6.4, the average spectral efficiency is presented with the same simulation parameters for the case where the CFO is compensated for the RS → MS link perfectly as presented in Equation (6.3). When the performance presented in Figures 6.3 and 6.4 are compared, it is seen that compensating the CFO for RS → MS link instead of BS → MS link improves performance when $SNR_{RD} > SNR_{SD}$. For example, when the CFO is compensated for the BS → MS link perfectly, then cooperative transmit diversity-1 is beneficial over cooperative receive diversity for CFOs up to 2.2% of the sub-carrier spacing (see Figure 6.3) and up to 3.45% (see Figure 6.4) when the CFO is compensated for the RS → MS link instead. This improvement becomes more significant as the SNR in the RS → MS link improves. For example, Figures 6.5 and 6.6 show the case where $SNR_{RD} = 9.5$ dB. When the CFO is compensated for the BS → MS link perfectly, then cooperative transmit diversity-1 is beneficial over cooperative receive diversity for CFOs up to 1.1% of the sub-carrier spacing (refer to Figure 6.5) and up to 2.94% of the sub-carrier spacing (refer to Figure 6.6) when the CFO is compensated for the RS → MS link instead.

When the SNR in the BS → RS link is very good, e.g., 48.5 dB as in the figures presented in this section, both the AF and DF-based relaying schemes perform the same when $\omega_{BS} - \omega_{RS} \neq 0$. As the results in this section show, if there is a large uncompensated residual CFO, the cooperative diversity schemes which use simultaneous transmissions from the RS and the BS are

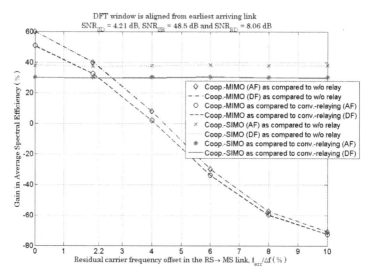

Figure 6.3 Spectral efficiency gain of cooperative transmit diversity-1 and cooperative receive diversity schemes as compared to w/o relay and conventional relaying with $SNR_{RD} = 8$ dB, $SNR_{SD} = 4.21$ dB and $SNR_{SR} = 48.5$ dB. The CFO is compensated for the $BS \rightarrow MS$ link perfectly.

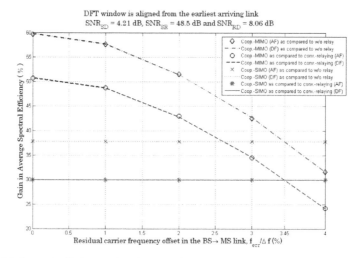

Figure 6.4 Spectral efficiency gain of cooperative transmit diversity-1 and cooperative receive diversity schemes as compared to w/o relay and conventional relaying with $SNR_{RD} = 8$ dB, $SNR_{SD} = 4.21$ dB and $SNR_{SR} = 48.5$ dB. The CFO is compensated for the $RS \rightarrow MS$ link perfectly.

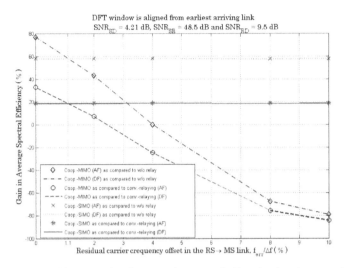

Figure 6.5 Spectral efficiency gain of cooperative transmit diversity-1 and cooperative receive diversity schemes as compared to w/o relay and conventional relaying with $SNR_{RD} = 9.5$ dB, $SNR_{SD} = 4.21$ dB and $SNR_{SR} = 48.5$ dB. The CFO is compensated for the BS $\rightarrow$ MS link perfectly.

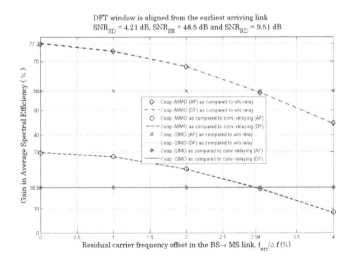

Figure 6.6 Spectral efficiency gain of cooperative transmit diversity-1 and cooperative receive diversity schemes as compared to w/o relay and conventional relaying with $SNR_{RD} = 9.5$ dB, $SNR_{SD} = 4.21$ dB and $SNR_{SR} = 48.5$ dB. The CFO is compensated for the RS $\rightarrow$ MS link perfectly.

no longer advantageous over the cooperative diversity schemes where only the RS transmit in the second phase. Therefore, efficient algorithms should be developed to remedy the effects of residual CFO in wireless relay networks. In summary, the optimal CFO compensation algorithm for wireless relay networks that uses simultaneous transmissions from the BS and RS should consider the ratio of $\varphi = \text{SNR}_{RD}/\text{SNR}_{SD}$. The design of such algorithm is interesting work for the future.

6.1.4.3 Effects of Time Offset on the Performance

In this paragraph, the effect of the time offset on the performance of various cooperative diversity schemes is presented. In Figure 6.7 the gain in average spectral efficiency in % is presented versus the time offset Δ which is given in terms of number of samples. It is assumed that the CFOs can be compensated perfectly and the FFT window is aligned from the earliest arriving link as visualized in Figure 6.1. In Figure 6.7, the considered average SNR values are $\text{SNR}_{SD} = 4.21$ dB and $\text{SNR}_{RD} = 8$ dB. When the time offset Δ is positive, the ISI is coupled from the BS $\rightarrow$ MS link if $\Delta + \tau_{\max} > $ CP. The term $\tau_{\max}$ in terms of number of samples represents the length of the channel impulse response of the BS $\rightarrow$ MS link. When the time offset Δ is negative, the ISI is coupled from the RS $\rightarrow$ MS link if $|\Delta| + \tau_{\max} > $ CP, where $\tau_{\max}$ in terms of number of samples represents the length of the channel impulse response of the RS $\rightarrow$ MS link. Over the region where relaying improves the performance, the RS $\rightarrow$ MS link has a greater SNR than that of the BS $\rightarrow$ MS link. Consequently, the degradation due to time offset over this region is steeper for $\Delta < 0$.

Hence, to reduce the ISI caused by the time offset problem, the position of the FFT window at the MS should be optimized for wireless relay networks. To determine the optimal FFT window positioning , the ratio of $\varphi = \text{SNR}_{RD}/\text{SNR}_{SD}$ should be considered. In the following, a possible FFT window positioning is proposed to further improve the tolerance to the time offset. This proposal is presented in Figure 6.8.

When $\Delta < 0$ and $|\Delta| + \tau_{\max} > $ CP and $\varphi > 1$, then the FFT window should be aligned from the earliest arriving link (i.e., from the BS $\rightarrow$ MS link) with a shift, i.e., μ samples. For the above condition, if the FFT window is not shifted then all the ISI samples will be coupled from the RS $\rightarrow$ MS link which causes higher degradation than the case where some part of the ISI is coupled from the BS $\rightarrow$ MS link. For the same condition, if the FFT window is shifted μ samples, then some part of the ISI will be coupled from the BS $\rightarrow$ MS link instead of all of the ISI is coupled from the RS $\rightarrow$

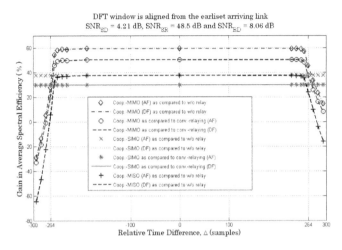

Figure 6.7 Spectral efficiency gain of cooperative transmit diversity-1, cooperative receive diversity and cooperative transmit diversity-2 schemes as compared to w/o relay and conventional relaying with $\text{SNR}_{RD} = 8$ dB, $\text{SNR}_{SD} = 4.2$ dB and $\text{SNR}_{SR} = 48.5$ dB. FFT window is aligned from the earliest arriving link.

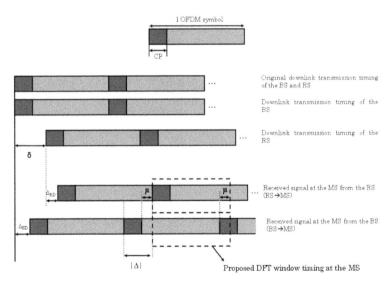

Figure 6.8 An example for the proposed FFT window timing at the MS when $\Delta < 0$ and $|\Delta| + \tau_{\text{max}} > \text{CP}$.

MS link. This will reduce the degradation on the performance of cooperative transmit diversity-1 and cooperative transmit diversity-2 schemes. The shift in FFT window timing, μ should basically be larger than shown in Figure 6.8 to further relieve the ISI coupled from the $RS \rightarrow MS$ link due to multi-path propagation. The optimization of the shift μ is an interesting future work.

6.2 Hardware Complexity Analysis of Cooperative Diversity Implementations at the MS

In order to achieve the post-processing SINR gain as compared to conventional relaying, signal combining should be used at the destination node. Combining methods differ in computational complexity depending on the realized cooperative diversity scheme. This results in varying processing time and hardware resource utilization at the MS. In order to evaluate the hardware requirements to achieve the cooperative diversity schemes presented in Chapter 3, the hardware-related measures using Field Programmable Gate Array (FPGA)-based implementation models have been analyzed in this section. The focus of this analysis is on the digital baseband processing blocks used at the MS. In IEEE 802.16e-based systems, the highest aggregate throughput in the system is achieved by using the transmission mode with 2048 sub-carriers. This is the most demanding case for the digital base-band processors used in the IEEE 802.16e-based systems. The results presented in this section are provided for this most demanding case. This way, the conclusions drawn can give insights to the practical feasibility of less demanding cases as well.

Figure 6.9 shows a comparison of FPGA elementary block usage and circuit power consumption estimates at the MS for different cooperative diversity scheme implementations. The considered cooperative diversity schemes are cooperative transmit diversity-1 and 2, cooperative receive diversity and cooperative selection diversity which were presented in Chapters 3 and 4. Parallel processing in the form of fast complex matrix multiplication is used to reduce the symbol processing time. This property is extremely important in real-time systems with large number of processed sub-carriers, where increase in symbol processing time can lead to infeasibility of the real-time operation. The number of DSP blocks on a single FPGA chip is significantly smaller as compared to other resources. Hence, $10\times$ scaling is used in the figure for the DSP block utilization measure. This is introduced to maintain a similar scale for all presented results in both visual

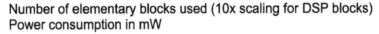

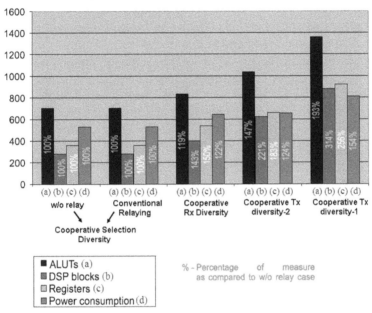

Figure 6.9 Hardware resource usage and power consumption estimates at the MS for the implementation of various cooperative diversity schemes.

and hardware-cost sense. Analysis shows that cooperative diversity schemes which require coherent signal combining introduce increased implementation cost at the MS in terms of hardware complexity and power consumption. To achieve coherent signal combining, the cooperative receive diversity requires the least number of complex multiplications. Cooperative diversity achieved with a cooperative-MIMO scheme (i.e., cooperative transmit diversity-1) requires the highest number of complex multiplications. Hence, as expected, the same order applies to the hardware resource usage. The evaluation presented in Figure 6.9 shows that conventional relaying does not bring additional processing complexity as compared to that of w/o relay scheme. Therefore, hardware complexity of the cooperative selection diversity is the same as that of w/o relay scheme. Furthermore, it is shown in Chapter 4 that cooperative selection diversity is promising in terms of end-to-end throughput performance with fixed relay stations deployed at strategic positions in the cell. Therefore, for wireless relay networks with fixed relay stations, cooperative selection diversity is a promising cooperative diversity scheme not only due to its end-to-end throughput performance, but also due to its simplicity in implementation at the MS. Power consumption estimates are related mainly to average processing workload which can be expressed as the number of arithmetical operations performed in a unit of time. As in the case of hardware resource usage, the w/o relay and conventional relaying schemes also present the lowest power consumption measures while cooperative transmit diversity-1 scheme introduces the highest power consumption. For cooperative receive diversity and cooperative transmit diversity-2 schemes, the power consumption estimates do not differ significantly from each other. The analysis presented in this section does not consider the idle phases at the MS which occur in the first phase of conventional relaying and cooperative transmit diversity-2. If this feature is taken into consideration, lower power consumption estimates will ensue for these schemes. This brings additional advantage as compared to other schemes.

The observations presented in this section suggest using cooperative diversity schemes which necessitate coherent signal combining at the MS only when the system performance gain is significant. Such significant performance gain can be observed for the users with high speeds where the use of instantaneous CSI at the BS may not be feasible to achieve cooperative selection diversity. For such users, this is due to higher time variations in the channel.

6.3 Conclusions and Future Work

In this chapter, the synchronization issues for OFDM(A)-based wireless relay networks have been presented. The time and carrier frequency offset issues have been analyzed. It is shown that both the time and frequency offset are not problematic with infrastructure-based RSs. The time offset problem can be overcome if the relays introduce an appropriate delay, either positive or negative, to their transmissions. To reduce the time offset and the ISI, the duration of this delay should be adjusted based on the position of the relay relative to the BS and the transmit power difference between the BS and the RS. This delay needs to be optimized. Such optimization is an interesting future study. Preamble transmission from the relays can further provide robustness in synchronization in a wireless relay network. As a future work, the evaluations on the cell throughput while considering the time offset problem can be investigated in a multiple relay scenario. This chapter further proposes an FFT window timing at the MSs which can relieve the effects of ISI when the time offset is large. It is shown in this chapter that, when the CFO between an RS and the BS is zero, then the CFO problem in OFDM(A)-based two-hop wireless relay networks resorts to that of single hop cellular networks. With infrastructure-based relays deployed at strategic positions in the cell, the relays can align their carrier frequency with that of BS at a high accuracy. This is thanks to the achievement of high SINR in the BS $\rightarrow$ RS links, the fixed (i.e., non-mobile) nature of the relays and the better hardware structure of an infrastructure-based RS as compared to that of a MS.

This chapter further analyzed the hardware complexity of implementing various cooperative diversity schemes at the MS. The investigations show that cooperative diversity schemes which require coherent signal combining at the MS result in significant complexity as compared to cooperative selection diversity. Therefore, for infrastructure-based wireless relay networks, the cooperative selection diversity is promising not only in terms of end-to-end throughput performance (as shown in Chapter 4) but also in terms of its simplicity in implementation at the MSs.

7

Conclusions and Future Work

In this book, various link adaptation mechanisms have been developed for OFDM(A)-TDD-based two-hop cellular networks with fixed relays. The systems investigated operate based on IEEE 802.16e and IEEE 802.16j standards. The DL transmissions to low mobility users have been considered with centralized control done by the BS. Various implementation issues have been identified and investigated for such networks.

In this book, an end-to-end link adaptation and selection method for wireless relay networks has been proposed. A frame structure in order to enable this proposal in OFDM(A)-TDD-based cellular wireless relay networks has been developed. Simple and efficient AMC decision rules have been developed for wireless relay networks. Such rules are able to take into account the fading conditions in all the wireless links constituting a relay network whereby the end-to-end throughput is optimized. The investigations show that, the end-to-end throughput performance with the proposed link adaptation and selection method is better than or equal to that of (1) w/o relay transmissions and (2) fixed relaying. The investigations further show that, with proper AMC algorithm, the AF-based relaying cannot outperform simple-AdDF-based relaying over the region where relaying improves the end-to-end throughput as compared to w/o relay transmissions. Hence, transmissions with the DF scheme is promising as the error propagation can be avoided by error detection techniques which are already inherent in wireless networks operating based on IEEE 802.16 standards. These conclusions further agree that drawn from information theoretic analysis presented in this book. In this book, a scheduler for implementation in two-hop cellular networks has been developed. This scheduler is able to consider all the instantaneous SINR conditions and the end-to-end throughput in order to schedule the users on the radio resources. It guarantees that the system level performance is always better than that of a w/o relay system and provides multi-user diversity. With this scheduler and multiple relays efficiently de-

137

ployed in the cell, the system level performance of various cooperative diversity schemes has been investigated comparatively. When the instantaneous SINR knowledge is available at the transmitter, the investigations show that the cooperative selection diversity is promising due to the following reasons. First of all, from the system level throughput point of view, it can perform as good as more complex diversity schemes which require coherent signal combining at the destination. At a given sub-channel, it requires transmissions either from the BS or the RS and hence allows rate adaptive relaying. This reduces the interference in the system as well. These conclusions further agree that drawn from information theoretic analysis presented in this book.

In this book, the end-to-end maximum achievable rate of various relaying schemes has been analyzed. The performance comparisons are provided over the SINR region where relaying improves the performance as compared to that of w/o relay transmission. The analysis shows that the DF-based relaying provides significant gain over the AF-based relaying for various relaying schemes. The AF-based relaying on the other hand cannot provide significant gain as compared to the DF-based relaying. The analysis shows that cooperative-MIMO scheme can provide an end-to-end achievable rate gain as compared to other schemes such as cooperative-SIMO, cooperative-MISO, conventional relaying and w/o relay. However, this gain can be achieved when $\gamma_{RD,i} \gg \gamma_{SD,i}$ and $\gamma_{RD,i} \gg \gamma_{SR,i}$. When $\gamma_{SR,i} \gg \gamma_{SD,i}$, $\gamma_{SR,i} \gg \gamma_{RD,i}$, cooperative-MISO scheme can achieve a higher end-to-end maximum achievable rate as compared to that of other schemes. This is mainly due to the fact that it provides not only post-processing SINR gain, but also enables rate adaptive relaying. As far as $\gamma_{SR,i} \gg \gamma_{SD,i}$ and $\gamma_{RD,i} \gg \gamma_{SD,i}$, all the relaying schemes perform the same. For such channel conditions, conventional relaying scheme should be used as it is the least complex relaying scheme.

In this book, the inefficiency of having the same MAC-PDU size over all the links constituting a multi-hop wireless communication system is investigated. To remedy this inefficiency and to provide an end-to-end optimization, a hop adaptive MAC-PDU size optimization is proposed. This proposal is analyzed for two-hop wireless relay networks with infrastructure-based relays deployed at strategic locations in the cell. The performance metrics are overhead reduction and goodput gain provided by the hop adaptive MAC-PDU size optimization. When there are a total of three infrastructure-based relays, a single BS, and 23 low mobility users in a Mobile-WiMAX network with relay support, the system level evaluations show that the average gain in goodput is 13% and average gains in the header and CRC overheads in relayed links is 13.54%. The investigations show that the average goodput gain via MAC-PDU size optimization reduces with increasing number of relays in the cell. On the other hand, the proposal brings significant reductions in the header and CRC overheads even with increasing number of relays in the cell.

In this book, the synchronization issues for OFDM-based wireless relay networks have been presented. The time and carrier frequency offset issues have been analyzed. It is shown that both the time offset and frequency offset are not problematic for infrastructure-based relaying. The time offset problem can be overcome if the relays introduce an appropriate delay to their transmissions. The duration of this delay depends on the position of the relay relative to the BS. Furthermore, it depends on the transmit power difference of the BS and the RS terminals and the delay spread of the channel. A new method which can relieve the effects of ISI caused by the time offset problem has been proposed. This method proposes to optimize the location of the FFT window at the MS in a wireless relay network. Such optimization should consider the SNR difference of the BS $\rightarrow$ MS and RS $\rightarrow$ MS links, the relative time offset and the delay spread of the channel.

The complexity of various cooperative diversity implementations has been investigated for MSs. Such investigation is done via implementing various cooperative diversity schemes on FPGA. The analysis and such implementation results presented in this book show that, the processing needed at the MSs to achieve cooperative-multi-antenna channel benefits should be introduced only when there is throughput gain over conventional relaying and w/o relay. By this way, the battery consumption at the mobiles can be reduced.

7.1 Future Work

Future work out of this book includes, but is not limited to, the investigation of the following items:

1. The transmission and reception with imperfect and/or reduced channel state information.
2. The transmissions to the users with high speeds. For such users, the sub-channel allocation is done on frequency diverse sub-carriers to provide frequency diversity [3]. In this case, new lookup tables should be prepared with average SINR conditions as instantaneous SINR conditions cannot be taken into account due to high variations in the channel.

3. The reduction in power consumption at the MS by using the algorithms developed in this book which introduce the cooperative diversity only when necessary.
4. The transmissions with HARQ.
5. The transmissions in a multi-cell environment with sectoring.
6. The beamforming for the BS → RS links which will further strengthen the link quality in the first hop. For transmissions in the BS → RS links, the use of beamforming will allow to use higher-rate modulation modes beyond 64-QAM. It can therefore reduce the multiplexing loss with the use of rate adaptive relaying schemes.

Appendix: List of Publications Produced during Ph.D. Studies

The dissemination of the results obtained within this Ph.D. work has been communicated to the research community via Intellectual Property Rights (IPRs), international conferences, journals and presentations. The following list classifies the publications in relation to the respective chapters of the thesis:

- Journal Manuscripts Submitted for Peer Review

    1. Başak Can, Rath Vannithamby, Hyunjeong (Hannah) Lee, Ali Taha Koç, Hop Adaptive MAC-PDU Size Optimization for Infrastructure Based Wireless Relay Networks, submitted to *IEEE Transactions on Wireless Communications*, February 2008.
    2. Başak Can, Halim Yanikomeroglu, Furuzan Atay Onat, Elisabeth De Carvalho and Hiroyuki Yomo, Performance of Link Adaptive Cooperative Diversity Schemes with Fixed Relays for Mobile WiMAX, submitted to *IEEE Transactions on Wireless Communications*, December 2007.

- Peer Reviewed Journal Publications

    1. Başak Can, Maciej Portalski, Hugo Simon Denis Lebreton, Simone Frattasi, Himal A. Suraweera, Implementation Issues for OFDM Based Multi-hop Cellular Networks, *IEEE Communications Magazine*, Special Issue on Technologies in Multi-hop Cellular Networks, September 2007.
    2. Başak Can, Hiroyuki Yomo and Elisabeth De Carvalho Link Adaptation and Selection Method for OFDM Based Wireless Relay Networks, *Journal of Communications and Networks (JCN)*, Special issue on MIMO OFDM and Its Applications, Editors: Vahid Tarokh, Siavash M. Alamouti, Visa Koivunen, KiHo Kim, June 2007.

3. Frank Fitzek, Başak Can, Ramjee Prasad, Marcos Katz, Traffic Analysis and Video Quality Evaluation of Multiple Description Coded Video Services for Fourth Generation Wireless IP Networks, *International Journal on Wireless Personal Communications*, Special Issue, Volume 35, Issue 1–2, pp. 187–200, October 2005.

- Intellectual Property Rights (IPRs) Filed

  1. Başak Can (Inventor), Hiroyuki Yomo (Inventor), Elisabeth De Carvalho (Inventor), Sivanesan Kathiravetpillai (Inventor), Dong-Seek Park (Inventor), Eun-Taek Lim (Inventor), Method for Adaptive Relaying in Communication System, Korean Intellectual Property Office, Patent No. KR2006-0126233, 2006-12-12.
  2. Başak Can (Inventor), Hiroyuki Yomo (Inventor), Elisabeth De Carvalho (Inventor), Jin-Kyu Koo (Inventor), Su-Ryong Jeong (Inventor), Young-Kwon Cho (Inventor), Jae-Weon Cho (Inventor), Dong-Seek Park (Inventor), Hokyu Choi (Inventor), Hybrid Forwarding Apparatus and Method for Cooperative Relaying in an OFDM Network, Korean Intellectual Property Office, Patent No. KR2006-0033844, 2006-04-14.

- Peer Reviewed Conference Publications

  1. Başak Can, Rath Vannithamby, Hyunjeong Hannah Lee, Ali Taha Koç, MAC-PDU Size Optimization for OFDMA Modulated Wireless Relay Networks, *IEEE Globecom Conference*, New Orleans, LA, USA, 30 November–4 December 2008.
  2. Başak Can, Maciej Portalski, Yannick Le Moullec, Hardware Aspects of Fixed Relay Station Design for OFDM(A) Based Wireless Relay Networks, *21st Canadian Conference on Electrical and Computer Engineering*, organized by IEEE Canada, Niagara Falls, Canada, May 2008.
  3. Başak Can, Halim Yanikomeroglu, Furuzan Atay Onat, Elisabeth De Carvalho, Hiroyuki Yomo, Efficient Cooperative Diversity Schemes and Radio Resource Allocation for IEEE 802.16j, *IEEE Wireless Communications and Networking Conference (WCNC)*, Las Vegas, USA, 31 March-3 April 2008.
  4. Başak Can, Hiroyuki Yomo, Elisabeth De Carvalho, Hybrid Forwarding Scheme for Cooperative Relaying in OFDM Based Net-

works, *IEEE International Conference on Communications (ICC)*, İstanbul, Turkey, June 2006.

5. Simone Frattasi, Başak Can, Frank Fitzek, Ramjee Prasad, Cooperative Services for 4G, *Proceedings of the 14th IST Mobile & Wireless Communications Summit*, Dresden, Germany, June 2005.

6. Frank H.P. Fitzek, Başak Can, Huan Cong Nguyen and Muhammad Imadur Rahman, Ramjee Prasad, Changhoi Koo, Cross Layer Optimization of OFDM Systems for 4G Wireless Communications, *9th International OFDM Workshop (InOWo 2004)*, Dresden, Germany, September 2004.

7. Frank H.P. Fitzek, Başak Can, Ramjee Prasad, Markos Katz, Overhead and Quality Measurements for Multiple Description Coding for Video Services, *Wireless Personal Multimedia Communications (WPMC)*, Abano Terme, Italy, September 2004.

8. Frank H.P. Fitzek, Başak Can, Ramjee Prasad, Markos Katz, Traffic Analysis of Multiple Description Coding of Video Services over IP Networks, *Wireless Personal Multimedia Communications (WPMC)*, Abano Terme, Italy, September 2004.

9. Frank H.P. Fitzek, Başak Can, Ramjee Prasad, D.S. Park, Youngkwon Cho, Application of Multiple Description Coding in 4G Wireless Communication Systems, *World Wireless Research Forum (WWRF) 8bis*, China, February 2004.

- Supervised Master Theses

    1. Maciej Portalski, Hugo Simon Denis Lebreton, Implementational Aspects of Cooperative Transmission Schemes in OFDM Based Wireless Relay Networks, 9th Semester Applied Signal Processing and Implementation (ASPI) Project Report, 2007. Supervisors: Başak Can, Yannick Le Moullec and M. Imadur Rahman.

    2. Maciej Portalski, Hardware Aspects of Fixed Relay Station Design for OFDM(A) Based Wireless Relay Networks, 10th Semester ASPI Project Report, 2007. Supervisors: Başak Can and Yannick Le Moullec.

- Deliverables

Various deliverable publications from 2004 to 2007, related to Joint Advanced Development Enabling 4G (JADE) Project.

References

[1] B. Can, M. Portlaski and Y. Le Moullec, Hardware Aspects of Fixed Relay Station Design for OFDM(A) Based Wireless Relay Networks, in *Proceedings of the 21st Canadian Conference on Electrical and Computer Engineering*, IEEE Canada, May 2008.

[2] B. Can, M. Portalski, H.S. Lebreton, S. Frattasi and H. Suraweera, Implementation Issues for OFDM-Based Multihop Cellular Networks, *IEEE Communications Magazine, Technologies in Multi-Hop Cellular Networks*, September 2007.

[3] Wimax Forum, Mobile WiMAX – Part I: A Technical Overview and Performance Evaluation, WiMAX Forum White Paper, June 2006.

[4] H. Yaghoobi, Scalable OFDMA Physical Layer in IEEE 802.16 WirelessMAN, *Intel Technology Journal*, vol. 8, no. 3, August 2004.

[5] R.U. Nabar, H. Bölcskei and F.W. Kneubühler, Fading Relay Channels: Performance Limits and Space-Time Signal Design, *IEEE Journal on Selected Areas in Communications*, vol. 22, no. 6, pp. 1099–1109, August 2004.

[6] P. Herhold, E. Zimmermann and G. Fettweis, Cooperative Multi-hop Transmission in Wireless Networks, *Computer Networks, Elsevier*, vol. 49, no. 3, pp. 299–324, June 2005.

[7] J.S. Park, H.J. Lee and M. Kim, Technical Standardization Status and the Advanced Strategies of the Next Generation Mobile Communications, in *Proceedings of the 8th International Conference on Advanced Communication Technology (ICACT)*, vol. 1, pp. 884–887, February 2006.

[8] M.R. Kibria, V. Mirchandani and A. Jamalipour, A Consolidated Architecture for 4G/B3G Networks, in *Proceedings IEEE Wireless Communications and Networking Conference*, March, vol. 4, pp. 2406–2411, 2005.

[9] A.J. Paulraj, D.A. Gore, R.U. Nabar and H. Bölcksei, An Overview of MIMO Communications – A Key to Gigabit Wireless, *Proceedings of the IEEE*, vol. 92, no. 2, pp. 198–218, February 2004.

[10] A. Paulraj, R.U. Nabar and D. Gore, *Introduction to Space-Time Wireless Communications*, Cambridge University Press, 2003.

[11] S.M. Alamouti, A Simple Transmit Diversity Technique for Wireless Communications, *IEEE Journal on Selected Areas in Communications*, vol. 16, no. 8, pp. 1451–1458, October 1998.

[12] V. Tarokh, H. Jafarkhani and A.R. Calderbank, Space-Time Block Codes from Orthogonal Designs, *IEEE Transactions on Information Theory*, vol. 45, no. 5, pp. 1456–1467, July 1999.

[13] M. Stoytchev, H. Safar, A.L. Moustakas and S. Simon, Compact Antenna Arrays for MIMO Applications, *IEEE Antennas and Propagation Society International Symposium*, vol. 3, pp. 708–711, July 2001.

[14] A. Nosratinia, T.E. Hunter and A. Hedayat, Cooperative Communication in Wireless Networks, *IEEE Communications Magazine*, vol. 42, no. 10, pp. 74–80, October 2004.

[15] M. Dohler, A. Gkelias and H. Aghvami, A Resource Allocation Strategy for Distributed MIMO Multi-Hop Communication Systems, *IEEE Communications Letters*, vol. 8, no. 2, pp. 99–101, February 2004.

[16] J.N. Laneman, G.W. Wornell and D.N.C. Tse, An Efficient Protocol for Realizing Cooperative Diversity in Wireless Networks, in *Proceedings of IEEE International Symposium on Information Theory*, p. 294, June 2001.

[17] P. Herhold, E. Zimmermann and G. Fettweis, Relaying and Cooperation – A System Perspective, *Proceedings of the 13th IST Mobile & Wireless Communications Summit*, Lyon, France, 27–30 June 2004.

[18] H. Hu, H. Yanıkömeroğlu, D. Falconer and S. Periyalwar, Range Extension without Capacity Penalty in Cellular Networks with Digital Fixed Relays, in *Proceedings of IEEE Global Telecommunications Conference*, vol. 5, pp. 3053–3057, December 2004.

[19] J.N. Laneman, Cooperative Diversity in Wireless Networks: Algorithms and Architectures, Ph.D. Dissertation, Massachusetts Institute of Technology, Department of Electrical Engineering and Computer Science, September 2002.

[20] R. Pabst, B. Walke, D. Schultz, P. Herhold, H. Yanikomeroglu, S. Mukherjee, H. Viswanathan, M. Lott, W. Zirwas, M. Dohler, H. Aghvami, D. Falconer and G. Fettweis, Relay Based Deployment Concepts for Wireless and Mobile Boradband Radio, *IEEE Communications Magazine*, vol. 42, pp. 80–89, September 2004.

[21] IEEE Computer Society and IEEE Microvawe Theory and Techniques Society, IEEE Standard for Local and metropolitan area networks Part 16: Air Interface for Fixed Broadband Wireless Access Systems, IEEE Technical Report, 2004.

[22] WiMAX Forum, WiMAX's Technology for LOS and NLOS Environments, WiMAX Forum White Papers, August 2004.

[23] IEEE 802.16's Relay Task Group. Available online: http://www.ieee802.org/16/relay/

[24] S. Frattasi and H. Fathi, A Cooperative ID for 4G, in *Cognitive and Cooperative Wireless Networks: Concepts, Methodologies and Visions*, Springer, 2007.

[25] M. Vajapeyam and U. Mitra, A Hybrid Space-Time Coding Scheme for Cooperative Networks, in *Proceedings of the 42nd Allerton Conference on Communications, Control and Computing*, October, 2004.

[26] N. Ahmed, M.A. Khojastepour and B. Aazhang, Outage Minimization and Optimal Power Control for the Fading Relay Channel, *IEEE Information Theory Workshop*, 24–29 October, pp. 458–462, 2004.

[27] R.U. Nabar and H. Bölcksei, Space-Time Signal Design for Fading Relay Channels, in *Proceedings of IEEE Global Telecommunications Conference*, 1–5 December, vol. 4, pp. 1952–1956, 2003.

[28] B. Can, H. Yomo and E. De Carvalho, Hybrid Forwarding Scheme for Cooperative Relaying in OFDM Based Networks, in *Proceedings of IEEE International Conference on Communications (ICC)*, İstanbul, Turkey, June 2006.

[29] B. Can (Inventor), H. Yomo (Inventor), E. De Carvalho (Inventor), J.K. Koo (Inventor), S.R. Jeong (Inventor), Y. K. Cho (Inventor), J. W. Cho (Inventor), D.S. Park

(Inventor) and H. Choi (Inventor), Hybrid Forwarding Apparatus and Method for Cooperative Relaying in an OFDM Network, Korean Intellectual Property Office, No. KR2006-0033844, April 2006.

[30] J.N. Laneman, D.N.C. Tse and G.W. Wornell, Cooperative Diversity in Wireless Networks: Efficient Protocols and Outage Behavior, *IEEE Transactions on Information Theory*, vol. 50, no. 12, pp. 3062–3080, December 2004.

[31] P. Herhold, E. Zimmermann and G. Fettweis, A Simple Cooperative Extension to Wireless Relaying, in *Proceedings of International Zürich Seminar on Communications*, Zürich, February, pp. 36–39, 2004.

[32] WiMAX Forum Member Organizations, WiMAX System Evaluation Methodology, WiMAX Forum, Technical Report, September 2007.

[33] The Relay Task Group of IEEE 802.16, Draft Standard for Local and Metropolitan Area Networks: Part 16, Air Interface for Fixed and Mobile Broadband Wireless Access Systems, Multihop Relay Specification, IEEE, Technical Report, August 2007.

[34] S.T. Chung and A.J. Goldsmith, Degrees of Freedom in Adaptive Modulation: A Unified View, *IEEE Transactions on Communications*, vol. 49, no. 9, pp. 1561–1571, September 2001.

[35] M. Asa, N. Natarajan, R. Peterson, S. Ramachandran and D. Chen, Scope Considerations for Mobile Multihop Relay, IEEE C80216mmr-05/013, September 2005.

[36] F. Chu and K. Chen, Fair Adaptive Radio Resource Allocation of Mobile OFDMA, in *Proceedings of IEEE International Symposium on PIMRC*, September, pp. 1–5, 2006.

[37] B. Sklar, Rayleigh Fading Channels in Mobile Digital Communication Systems, Part I: Characterization, *IEEE Communications Magazine*, July 1997.

[38] S. Haykin, *Communication Systems*, John Wiley & Sons, 1994.

[39] J.G. Proakis, *Digital Communications*, McGraw-Hill, 2001.

[40] IEEE 802.16 Broadband Wireless Access Working Group, Channel Models for Fixed Wireless Applications, IEEE, Technical Report, June 2003.

[41] D.S. Baum, J. Hansen, J. Salo, G. Del Galdo, M. Milojevic and P. Kyösti, An Interim Channel Model for Beyond-3G Systems: Extending the 3GPP Spatial Channel Model (SCM), in *Proceedings of IEEE VTC, Spring*, vol. 5, pp. 3132–3136, 2005.

[42] B. Can, H. Yomo and E. De Carvalho, Link Adaptation and Selection Method for OFDM Based Wireless Relay Networks, *Journal of Communications and Networks*, Special issue on MIMO-OFDM and Its Applications, June 2007. Online available at http://kom.aau.dk/~bc/

[43] Z. Lin, E. Erkip and M. Ghosh, Adaptive Modulation for Coded Cooperative Systems, in *Proceedings of IEEE 6th Workshop on Signal Processing Advances in Wireless Communications*, June, pp. 615–619, 2005.

[44] K.A. Witzke and C. Leung, A Comparison of Some Error Detecting CRC Code Standards, *IEEE Transactions on Communications*, vol. 33, no. 9, September 1985.

[45] IEEE, IEEE Standard for Local and Metropolitan Area Networks Part 16: Air Interface for Fixed and Mobile Broadband Wireless Access Systems Amendment 2: Physical and Medium Access Control Layers for Combined Fixed and Mobile Operation in Licensed Bands and Corrigendum 1, IEEE, Technical Report, 2006. Online available at http://www.ieeexplore.ieee.org/xpl/standardstoc.jsp?isnumber=33683&isYe%ar=

[46] Ö. Oyman, N. Laneman and S. Sandhu, Multihop Relaying for Broadband Wireless Mesh Networks: From Theory to Practice, *IEEE Communications Magazine*, pp. 116–122, November 2007.

[47] T.S. Rappaport, *Wireless Communications*, Prentice Hall Communications Engineering and Emerging Technologies Series, Prentice Hall, 2002.

[48] V. Erceg, L.J. Greenstein, S.Y. Tjandra, S.R. Parkoff, A. Gupta, B. Kulic, A.A. Julius and R. Bianchi, An Empirically Based Path Loss Model for Wireless Channels in Suburban Environments, *IEEE Journal on Selected Areas in Communications*, vol. 17, no. 7, pp. 1205–1211, July 1999.

[49] R. Van Nee and R. Prasad, *OFDM for Wireless Multimedia Communications*, 2000, 46 pp.

[50] B. Can (Inventor), H. Yomo (Inventor), E. De Carvalho (Inventor), S. Kathiravetpillai (Inventor), D.S. Park (Inventor) and E.T. Lim (Inventor), Method for Adaptive Relaying in Communication System, Korean Intellectual Property Office, No. KR2006-0126233, December 2006.

[51] B. Can, H. Yanikomeroglu, F. Atay Onat, E. De Carvalho and H. Yomo, Efficient Cooperative Diversity Schemes and Radio Resource Allocation for IEEE 802.16j, in *Proceedings of IEEE Wireless Communications and Networking Conference (WCNC)*, 31 March–3 April, 2008.

[52] M. Hu and J. Zhang, Opportunistic Multi-Access: Multiuser Diversity, Relay-Aided Opportunistic Scheduling and Traffic-Aided Smooth Admission, *Mobile Networks and Applications*, vol. 9, no. 4, 2004.

[53] G. Li and H. Liu, Resource Allocation for OFDMA Relay Networks with Fairness Constraints, *IEEE Journal on Selected Areas in Communications*, vol. 24, no. 11, November 2006.

[54] I. Hammerstrom, M. Kuhn and A. Wittneben, Channel Adaptive Scheduling for Cooperative Relay Networks, in *Proceedings of IEEE VTC*, Fall, vol. 4, pp. 2784–2788, 2004.

[55] F. Atay Onat, A. Adinoyi, Y. Fan, H. Yanıkömeroğlu and J.S. Thompson, Optimum Threshold for SNR-based Selective Digital Relaying Schemes in Cooperative Wireless Networks, in *Proceedings of IEEE Wireless Communications and Networking Conference (WCNC)*, March 2007.

[56] M. Kaneko and P. Popovski, Adaptive Resource Allocation in Cellular OFDMA System with Multiple Relay Stations, in *Proceedings of IEEE Vehicular Technology Conference (VTC)*, Spring, pp. 3026–3030, 2007.

[57] M. Kaneko and P. Popovski, Radio Resource Allocation Algorithm for Relay-aided Cellular OFDMA System, in *Proceedings of IEEE International Conference on Communications*, June 2007.

[58] M. Asa, D. T. Chen and N. Natarajan, Relay Strategy of Broadcast Messages in Mobile Multihop Relay, IEEE 802.16 Presentation Submission Template (Rev. 8.3), No. IEEE C802.16mmr-06/008, January 2006.

[59] T. Park, O. Shin and K. B. Lee, Proportional Fair Scheduling for Wireless Communication with Multiple Transmit and Receive Antennas, in *Proceedings of IEEE VTC*, Fall, vol. 3, pp. 1573–1577, 2003.

[60] G. Li and H. Liu, On the Capacity of Broadband Relay Networks, in *Proceedings of Asilomar Conference on Signals, Systems and Computers*, 7–10 November, Pacific Grove, CA, 2004.

[61] T.M. Cover and J.A. Thomas, in *Elements of Information Theory*, D.L. Schilling (Ed.), Wiley Series in Telecommunications, Wiley, New York, 1991.

[62] E. Telatar, Capacity of Multi-Antenna Gaussian Channels, *European Transactions on Telecommunications*, vol. 10, pp. 585–595, 1999.

[63] T.S. Chu and L.J. Greenstein, A Quantification of Link Budget Differences between the Cellular and PCS Bands, *IEEE Transactions in Vehicular Technology*, vol. 48, no. 1, pp. 60–65, January 1999.

[64] W.C. Jakes and D.O. Reudink, Comparison of Mobile Radio Transmission at UHF and X-Band, *IEEE Transactions in Vehicular Technology*, pp. 10–13, October 1967.

[65] Y. Okumura, E. Ohmori, T. Kawano and K. Fukua, Field Strength and Its Variability in UHF and VHF Land-Mobile Radio Service, *Rev. Elec. Commun. Lab.*, vol. 16, no. 9, 1968.

[66] R. You, H. Li and Y. Bar-Ness, Diversity Combining with Imperfect Channel Estimation, *IEEE Transactions on Communications*, vol. 53, no. 10, pp. 1655–1662, October 2005.

[67] T. Pollet, M. Van Bladel and M. Moeneclaey, BER Sensitivity of OFDM Systems to Carrier Frequency Offset and Wiener Phase Noise, *IEEE Transactions on Communications*, vol. 43, pp. 191–193, 1995.

Index

151

Curriculum Vitae

Başak Can was born in Ceyhan/Adana, Turkey. She received the Electrical and Electronics Engineer degree from Ege University, Turkey in 2001. She received the Master of Science degree in Telecommunications from Bilkent University, Turkey in 2003. In 2004, she joined the Department of Electronic Systems of Aalborg University where she conducted her Ph.D. studies in wireless communications. On 20 May 2008, she was awarded the Degree of Doctor of Philosophy. Within her Ph.D. studies she has been working on Joint Advanced Development Enabling 4G (JADE) project in cooperation with Samsung Electronics Co. Ltd., Republic of Korea. From mid January till the end of May 2007 she was a visiting researcher at the Department of Systems and Computer Engineering of Carleton University, Canada. Since August 2007, she has been working at Intel Corporation, U.S.A. She took part in the development of the next generation mobile WiMAX standard and the radio conformance tests for Wimax products. Her main research and development interests are cellular wireless networks, digital signal processing/processors, OFDMA, MIMO techniques, cross-layer optimization, emerging wireless transmission standards and products.

For Product Safety Concerns and Information please contact our EU
representative GPSR@taylorandfrancis.com
Taylor & Francis Verlag GmbH, Kaufingerstraße 24, 80331 München, Germany

www.ingramcontent.com/pod-product-compliance
Ingram Content Group UK Ltd.
Pitfield, Milton Keynes, MK11 3LW, UK
UKHW021121180425
457613UK00005B/182